Nicolas Brandes

Oxidative Thiol Modifications in Pro- and Eukaryotic Organisms

Nicolas Brandes

Oxidative Thiol Modifications in Pro- and Eukaryotic Organisms

The Influence of Oxidative and Nitrosative Stress on Survival and Aging of Escherichia coli and Saccharomyces cerevisiae

Südwestdeutscher Verlag für Hochschulschriften

Impressum / Imprint
Bibliografische Information der Deutschen Nationalbibliothek: Die Deutsche Nationalbibliothek verzeichnet diese Publikation in der Deutschen Nationalbibliografie; detaillierte bibliografische Daten sind im Internet über http://dnb.d-nb.de abrufbar.
Alle in diesem Buch genannten Marken und Produktnamen unterliegen warenzeichen-, marken- oder patentrechtlichem Schutz bzw. sind Warenzeichen oder eingetragene Warenzeichen der jeweiligen Inhaber. Die Wiedergabe von Marken, Produktnamen, Gebrauchsnamen, Handelsnamen, Warenbezeichnungen u.s.w. in diesem Werk berechtigt auch ohne besondere Kennzeichnung nicht zu der Annahme, dass solche Namen im Sinne der Warenzeichen- und Markenschutzgesetzgebung als frei zu betrachten wären und daher von jedermann benutzt werden dürften.

Bibliographic information published by the Deutsche Nationalbibliothek: The Deutsche Nationalbibliothek lists this publication in the Deutsche Nationalbibliografie; detailed bibliographic data are available in the Internet at http://dnb.d-nb.de.
Any brand names and product names mentioned in this book are subject to trademark, brand or patent protection and are trademarks or registered trademarks of their respective holders. The use of brand names, product names, common names, trade names, product descriptions etc. even without a particular marking in this work is in no way to be construed to mean that such names may be regarded as unrestricted in respect of trademark and brand protection legislation and could thus be used by anyone.

Verlag / Publisher:
Südwestdeutscher Verlag für Hochschulschriften
ist ein Imprint der / is a trademark of
OmniScriptum GmbH & Co. KG
Heinrich-Böcking-Str. 6-8, 66121 Saarbrücken, Deutschland / Germany
Email: info@svh-verlag.de

Herstellung: siehe letzte Seite /
Printed at: see last page
ISBN: 978-3-8381-1611-2

Zugl. / Approved by: Würzburg, Universität, Diss., 2010

Copyright © 2010 OmniScriptum GmbH & Co. KG
Alle Rechte vorbehalten. / All rights reserved. Saarbrücken 2010

Table of Contents

Table of Contents I
Abbreviations VII

Summary 1

I. Summary 1

II. Zusammenfassung 5

1 Introduction 9

1.1 Reactive oxygen species and oxidative stress 10

1.2 Cysteine thiols — Central components of redox-sensitive nanoswitches 11

1.3 Changing gene transcription to protect cells against oxidative stress 14

1.4 Yap1 — Changing the redox state means switching homes 14

1.4.1 Yap1 — Member of a two-component redox relay 15

1.4.2 Redox-mediated fine-tuning of Yap1's functional activity 17

1.4.3 Activation of Yap1 — More than one way to get the response 18

1.4.4 Yap1 — A prototype for emerging concepts in redox regulation? ... 19

1.4.5 Redox theme with variations — The Nrf2-Keap1 connection 19

1.5 GapDH — Putting cellular metabolism under redox-control 20

1.5.1 Re-Routing the metabolic flux to protect cells against oxidative damage 22

1.5.2 ROS-mediated GapDH aggregation and its role in apoptosis 23

1.5.3 Role of GapDH's redox-regulation in signaling 25

II   Table of Contents

    1.5.4 GapDH — One of many redox-regulated metabolic enzymes 27

1.6 **PTP1B — Redox regulation of eukaryotic signal transduction cascades** 27

    1.6.1 PTP1B — Using cyclic sulfenamide formation as redox switch 28

    1.6.2 Overoxidation of PTP1B — More than a dead-end product? 30

    1.6.3 Other redox-regulated signal transduction cascades in eukaryotes 31

1.7 **Concluding remarks** 32

2 Nitrosative Stress Treatment of *E. coli* Targets Distinct Set of Thiol-containing Proteins 35

2.1 **Introduction** 36

2.2 **Material and Methods** 37

    2.2.1 Bacterial strains and culture conditions 37

    2.2.2 Plasmid construction 38

    2.2.3 Differential thiol trapping 38

        2.2.3.1 Sample preparation and differential thiol trapping 38

        2.2.3.2 Staining of the gels, storage phosphor autoradiography, and image analysis 39

        2.2.3.3 Data analysis and identification of proteins from 2D gels 40

    2.2.4 Purification of Glutamate Synthase (GOGAT) 40

    2.2.5 Enzyme assays 41

    2.2.6 UV absorbance and analysis of GOGAT's thiol status using mass spectrometry 41

2.3 **Results and Discussion** 42

    2.3.1 Global analysis of the thiol-disulfide state of *E. coli* proteins upon DEA/NO treatment 42

    2.3.2 NO-treatment causes oxidative thiol modification in a distinct set of *E. coli* proteins 46

    2.3.3 Identification of essential RNS-sensitive proteins involved in DEA/NO-induced growth inhibition 48

- 2.3.4 Activity of IlvC is affected by DEA/NO treatment *in vivo* 50
- 2.3.5 Glutamate Synthase — A protein specifically sensitive to nitrosative stress 51
- 2.3.6 Nitrosative stress treatment causes [4Fe-4S] cluster disassembly of GOGAT *in vitro* 54
- 2.3.7 RNS targets cysteines of GOGAT's [Fe-S] cluster 55

2.4 **Conclusions** 58

3 Oxidation-sensitive yeast proteins in subcellular compartments 61

3.1 **Introduction** 62

3.2 **Material and Methods** 65

- 3.2.1 Strains and cell growth 65
- 3.2.2 Analysis of glucose concentration 65
- 3.2.3 Purification of subcellular compartments 65
  - 3.2.3.1 Purification of mitochondria 65
  - 3.2.3.2 Purification of vacuoles 66
  - 3.2.3.3 Purification of nuclei 67
- 3.2.4 Differential thiol trapping of whole cells using ICAT 68
  - 3.2.4.1 Differential thiol trapping procedure 68
  - 3.2.4.2 Tryptic digest and purification of ICAT-labeled peptides 68
- 3.2.5 LC-MS/MS analysis 69
- 3.2.6 Data analysis 70
- 3.2.7 Design of the OxICAT program by extending msInspect 71
  - 3.2.7.1 File conversion 71
  - 3.2.7.2 Detection of peptides 71
  - 3.2.7.3 Retention time correction and peptide matching 71
  - 3.2.7.4 Detection of ICAT pairs 72
  - 3.2.7.5 Quantification of the oxidation state 72
  - 3.2.7.6 Quality control 72
- 3.2.8 Enhanced detection of organelle-specific yeast proteins using OxICAT spiking 73

3.2.9 Prediction of pK_a value, cysteine accessibility, secondary structure and fold propensity 73

3.3 Results and Discussion 74

    3.3.1 OxICAT as method to identify the redox state of proteins in yeast 74

    3.3.2 Analysis of the *in vivo* redox status of organelle-specific proteins in yeast 76

    3.3.3 The majority of proteins are in their reduced state during exponential growth 80

    3.3.4 Cu/Zn-Sod and Tim10 — Two highly oxidized proteins in yeast 82

    3.3.5 OxICAT to quantify the oxidation status of each cysteine separately in multi-cysteine proteins 86

    3.3.6 Identification of oxidation-sensitive yeast targets 86

    3.3.7 Translation, stress response and amino acid/carbohydrate metabolism are major targets of H_2O_2-mediated oxidation 88

    3.3.8 Peroxide-mediated regulation of major functional classes 89

        3.3.8.1 Carbohydrate metabolism 89

        3.3.8.2 Amino acid biosynthesis 91

        3.3.8.3 Protein translation 92

        3.3.8.4 Stress Response and Protein Folding 93

    3.3.9 Localization of H_2O_2-sensitive target proteins 94

    3.3.10 Structural analysis reveals no differences between H_2O_2-sensitive and insensitive cysteine residues 95

    3.3.11 Aconitase as major target of superoxide stress in yeast 98

3.4 Conclusions 101

3.5 Appendix Chapter 3 105

4 Collapse of the Cellular Redox Balance: A Key Event for Chronological Aging in Yeast? 119

    4.1 Introduction 119

4.2 Material and Methods 122

4.2.1 Strains and cell growth 122
4.2.2 Determination of cell viability 122
4.2.3 Differential thiol trapping of chronological aging yeast 122
4.2.4 Clustering analysis 123
4.2.5 Prediction of fold propensity 123
4.2.6 Analysis of glucose concentration 123
4.2.7 Determination of total cellular ATP levels 123
4.2.8 Determination of intracellular GSH, GSSG, and cysteine concentration and calculation of redox potentials 124

4.3 Results and Discussion 126

4.3.1 Glucose concentration — A determinant of chronological life span in yeast 126
4.3.2 Detection of oxidation-sensitive protein cysteines during chronological aging 127
4.3.3 Chronological aging in yeast — Collapse of the cellular redox homeostasis? 128
4.3.4 Reduction in glucose concentration causes delayed collapse of the general redox balance 129
4.3.5 Oxidation of the active site cysteines in GapDH illustrates the redox collapse during CLS 130
4.3.6 Clustering analysis reveals early targets of oxidation 132
4.3.7 Oxidation of thioredoxin reductase precedes general redox collapse 136
4.3.8 Other early targets of oxidation in chronologically aging yeast 140
4.3.8.1 ATP-dependent molecular chaperone Ccd48 140
4.3.8.2 Semialdehyde dehydrogenase Lys2 141
4.3.8.3 Pyruvate carboxylase Pyc2 141
4.3.8.4 T-complex protein 1 subunit delta Cct4 142
4.3.8.5 Early oxidation of proteins involved in protein translation 142
4.3.9 Proteins that maintain their oxidation status during chronological yeast aging 143
4.3.10 Early oxidation targets — Cysteines in unfolded regions 144

4.3.11 Metabolic and redox metabolomic changes in aging yeast cells 145

    4.3.11.1 Analysis of the glucose level in chronological aging yeast 145

    4.3.11.2 ATP level in chronological aging yeast 145

4.3.12 Analysis of intracellular GSH/GSSG and cysteine levels in chronological aging yeast 146

4.4 Conclusions 150

4.5 Appendix Chapter 4 155

REFERENCES 179

Abbreviations

0-9
2D	two-dimensional

A
A. thaliana	*Arabidopsis thaliana*
aa	amino acid
ACN	acetonitrile
AHP1	alkyl hydroperoxide reductase
AP-1	activator protein-1
ARE	antioxidant response element
ATF	activating transcription factor
ATP	adenosine 5'-triphosphate

B
bp	basepair
BSA	bovine serum albumin
bZIP	basic leucine zipper

C
C. elegans	*Caenorhabditis elegans*
CAT	Catalase
CFU	colony forming unit
CLS	chronological life span
CR	caloric restriction
CRD	cysteine-rich domain
CRM1	nuclear export receptor
CS	citrate synthase
CTAB	cetyl trimethyl ammonium bromide
CTWC	Coupled Two Way Clustering
Cu/Zn	copper/zinc
CYS	cysteine

D
Da, kDa	dalton, kilo dalton
DAB	denaturing alkylating buffer
DEA/NO	diethylamine NONOate
DNA	deoxyribonucleic acid
DTT	dithiothreitol

E
E. coli	*Escherichia coli*
ER	endoplasmic reticulum

G
g	gravitational force
GAP	glyeraldehyde-3-phosphate
GAPDH	glyceraldehyde-3-phosphate dehydrogenase
GFP	green fluorescent protein

GltB	glutamate synthase large subunit
GltD	glutamate synthase small subunit
GND	6-phosphogluconate dehydrogenase
GOGAT	glutamine:2-oxoglutarate amidotransferase (short: glutamate synthase)
GPX	glutathione peroxidase
GRX	glutaredoxin
GSH	glutathione (reduced)
GSNO	S-nitrosoglutathione
GSSG	glutathione disulfide

H

h	hours
$H2O_{dd}$	double distilled water
H_2O_2	hydrogen peroxide
HSF-1	heat shock transcription factor 1
HSP33	heat shock protein 33

I

IAM	iodoacetamide
ICAT	isotope coded affinity tag
IEF	isoelectric focusing
IlvC	ketol-acid reductoisomerase
IPTG	isopropyl thio-β-D-galactoside

K

Keap1	Kelch-like ECH-associated protein-1
LB	Luria-Bertani

L

LC	liquid chromatography

M

m, cm, µm, nm	meter, centimeter, micrometer, nanometer
M, mM, µM	molar, millimolar, micromolar
m/z	mass-to-charge ratio
MAPK	mitogen-activated protein kinase
MEV	Multi Experiment Viewer
min	minutes
MS	mass spectrometry
MS/MS	tandem mass spectrometry
mV	millivolt

N

NaDOC	sodium deoxycholate
NADPH	nicotinamide adenine dinucleotide phosphate
NEM	N-ethylmaleimide
NES	nuclear export signal
NF-1	nuclear factor 1
NF-κB	nuclear factor kappa B
NLS	nuclear localization signal

•NO	nitric oxide radical
NO⁻	nitroxyl anion
NO_2	nitrogen dioxide
N_2O_3	dinitrogen trioxide
Nrf2	nuclear factor erythroid-2 related factor
O	
$^1O_2$	singlet oxygen
•O_2^-	superoxide anion radical
•OH	hydroxyl radical
ONOO⁻	peroxynitrite
P	
PAGE	polyacrylamide gel electrophoresis
PBS	phosphate buffered saline
PCR	polymerase chain reaction
PDGF	platelet-derived growth factor
PDC	pyruvate dehydrogenase complex
PDI	protein disulfide isomerase
pI	isoelectric point
PI	propidium iodide
PI(3)K	phosphoinositide 3-kinase
PIC	protease inhibitor cocktail
PIP_3	PtdIns(3,4,5)P3(phosphatidylinositol(3,4,5)-trisphosphate
PMSF	phenylmethylsulphonyl fluoride
PPP	pentose phosphate pathway
PRX	peroxiredoxin
PTEN	phosphatase with sequence homology to tensin
PTP	protein tyrosine phosphatase
PTP1B	tyrosine-protein phosphatase non-receptor type 1
pTyr	phosphotyrosine
R	
RNS	reactive nitrogen species
ROS	reactive oxygen species
rpm	rounds per minute
S	
S. typhimurium	*Salmonella typhimurium*
SC	synthetic complete
SCD	synthetic complete dextrose
SCG	synthetic complete glycerol
SD	standard deviation
SDS	sodium dodecyl sulfate
SE	standard error
SGD	*S. cerevisiae* Genome Database
SO_2H	sulfinic acid
SO_3H	sulfonic acid
SOD	superoxide dismutase

SOH	sulfenic acid
SPC1	suppressor of phosphatase 2C
SRX	sulfiredoxin
T	
TCA	trichloroacetic acid
TCEP	Tris(2-carboxyethyl)phosphine hydrochloride
TC-PTP	T-Cell protein tyrosine phosphatase
TDH	glyceraldehyde-3-phosphate dehydrogenase isozyme
TFA	trifluoroacetic acid
TIM	mitochondrial import inner membrane translocase subunit
TPI	triosephosphate isomerase
TRR	thioredoxin reductase
TRX	thioredoxin
U	
USF	upstream stimulatory factor
UV	ultraviolet
V	
V	volt
v/v	volume per volume
W	
WT	wild type
wt/vol	weight per volume
Y	
YAP1	yeast AP-1
YBP1	Yap1 binding protein
YFP	yellow fluorescent protein
YPD	yeast peptone dextrose
YPG	yeast peptone glycerol

Summary

I. Summary

Cysteines play important roles in the biochemistry of many proteins. The high reactivity, redox properties, and ability of the free thiol group to coordinate metal ions designate cysteines as the amino acids of choice to form key catalytic components of many enzymes. Moreover, cysteines readily react with reactive oxygen and nitrogen species to form reversible oxidative thiol modifications. Over the last few years, an increasing number of proteins have been identified that use redox-mediated thiol modifications to modulate their function, activity, or localization. These redox-regulated proteins are central players in numerous important cellular processes, including translation, protein folding, stress protection, signal transduction, and apoptosis.

First aim of this study was to discover nitric oxide (NO) sensitive proteins in Escherichia coli (*E. coli*), whose redox-mediated functional changes might explain the physiological alterations observed in *E. coli* cells suffering from NO-stress. To identify *E. coli* proteins that undergo reversible thiol modifications upon NO-treatment *in vivo*, I applied a differential thiol trapping technique combined with two-dimensional gel analysis. Only a very select set of 10 proteins was found to contain thiol groups sensitive to NO-treatment. Subsequent genetic studies revealed that the oxidative modifications of two of these proteins (Dihydrolipoamide acetyltransferase and Ketol-acid reductoisomerase) are, in part, responsible for the observed NO-induced growth inhibition, indicating that RNS-mediated modifications play important physiological roles. Noteworthy, the majority of identified protein targets turned out to be specifically sensitive towards reactive nitrogen species. This oxidant specificity was tested on one NO-sensitive protein, the small subunit of glutamate synthase. *In vivo* and *in vitro* activity studies demonstrated that glutamate synthase rapidly inactivates upon nitric oxide treatment but is resistant towards other oxidative stressors. These results imply that reactive oxygen and nitrogen species affect distinct physiological processes in bacteria.

The second aim of my study was to identify redox-sensitive proteins in Saccharomyces cerevisiae (*S. cerevisiae*) and to use their redox state as *in vivo* read-out to assess onset, extent and role of oxidative stress during the eukaryotic aging process. I first determined the precise *in vivo* thiol status of almost 300 yeast proteins located in the cytosol and sub-

2   Summary

cellular compartments (*i.e.*, mitochondria, nuclei, vacuoles) of exponentially growing *S. cerevisiae* cells using a highly quantitative mass spectrometry based thiol trapping technique, called OxICAT. The identified proteins can be clustered in four groups: 1) proteins, whose cysteine residues are oxidation resistant; 2) proteins with structurally or functionally important cysteine modifications, most likely disulfide bonds; 3) proteins with highly oxidation-sensitive active site cysteines, which are partially oxidized in exponentially growing yeast cells due to their exquisite sensitivity towards low amounts of ROS; 4) proteins that are reduced in exponentially growing cells but harbor redox-sensitive cysteine(s) that affect the catalytic function of the protein during oxidative stress. These oxidative stress sensitive proteins were identified by short-term exposure of exponentially growing yeast cells to sublethal concentrations of H_2O_2 or superoxide. It was shown that the major targets of peroxide- and superoxide-mediated stress in the cell are proteins involved in translation, glycolysis, TCA cycle and amino acid biosynthesis. These targets indicate that cells rapidly redirect the metabolic flux and energy towards the pentose phosphate pathway in an attempt to ensure the production of the reducing equivalent NADPH to counterattack oxidative stress. These results reveal that the quantitative assessment of a protein's oxidation state is a valuable tool to identify catalytically active and redox-sensitive cysteine residues and illustrate how reversible posttranslational thiol modifications are used by the cells to coordinate a metabolic response, which protects against the effects of oxidative stress and assures cell survival.

The OxICAT technology was then used to precisely determine the extent and onset of oxidative stress in chronologically aging *S. cerevisiae* cells by utilizing the redox status of proteins as physiological read-out. I found that chronological aging yeast cells undergo a global collapse of the cellular redox homeostasis, which precedes cell death. The onset of this collapse appears to correlate with the yeast life span, as caloric restriction, defined by a reduction in nutrient availability without malnutrition, increases the life span and delays the redox collapse. These results suggest that maintenance of the redox balance might contribute to the life expanding benefits of regulating the caloric intake of yeast. Clustering analysis of all oxidatively modified proteins in chronological aging yeast revealed a subset of proteins whose oxidative thiol modifications significantly precede the general redox collapse. Oxidation of these early target proteins, which most likely results in a loss of their activity, might contribute to or even cause the observed loss of redox homeostasis (*i.e.*, thioredoxin reductase) or cell death (*i.e.*, cell division control protein 48) in chronologically aging yeast. These studies in aging yeast expand our understanding how changes in redox homeostasis and oxidative stress affect the life span of yeast cells and further confirm the importance of oxidative thiol modifications as one of the key posttranslational modifications in pro- and eukaryotic organisms.

II. Zusammenfassung

Cystein spielt eine wichtige Rolle in der Biochemie vieler Proteine. Aufgrund der Redox-Eigenschaften und der hohen Reaktivität der freien Thiol-Gruppe sowie dessen Fähigkeit Metallionen zu koordinieren, ist die Aminosäure Cystein oftmals Bestandteil von katalytischen Zentren vieler Enzyme. Zudem lassen sich Cysteine in Anwesenheit von reaktiven Sauerstoff- und Stickstoffspezies leicht reversibel oxidativ modifizieren. Innerhalb der letzten Jahre wurde zunehmends gezeigt, dass Proteine redox-bedingte Thiol-Modifikationen nutzen, um Veränderungen ihrer Funktion, Aktivität oder Zell-Lokalisierung zu steuern. Diese redox-regulierten Proteine spielen eine zentrale Rolle in vielen physiologischen Prozessen, wie z.b. Translation, Proteinfaltung, Schutz vor Stress, Signaltransduktion oder Apoptose.

Das erste Ziel meiner Arbeit war die Identifizierung von Stickstoffmonoxid (NO)-sensitiven Proteinen in *Escherichia coli (E. coli)*. Die redox-bedingten Funktions-änderungen solcher Proteine erklären möglicherweise die veränderte Physiologie von *E. coli* Zellen, die unter NO-Stress leiden. Um *E. coli* Proteine zu identifizieren, die unter Einwirkung von NO-Stress reversibel Thiol-modifiziert werden, wandte ich eine Kombination aus differentiellem Thiol-Trapping und 2D Gel-Elektrophorese an. Es wurde eine kleine Gruppe von zehn Proteinen identifiziert, welche NO-sensitive Thiol-Gruppen enthalten. Nachfolgende genetische Studien ergaben, dass die oxidativen Modifikationen an zwei dieser Proteine (*Dihydrolipoamide acetyltransferase* und *Ketolacid reductoisomerase*) mitverantwortlich sind für die NO-induzierte Wachstumshemmung. Dieses Ergebnis deutet darauf hin, dass durch reaktive Stickstoffspezies verursachte Modifikationen eine wichtige physiologische Rolle spielen. Bemerkenswert ist es, dass die Mehrheit der identifizierten Proteine speziell nur gegen reaktive Stickstoffspezies empfindlich ist. Diese spezifische Sensitivität gegenüber einem bestimmten Oxidationsmittel wurde an einem der identifizierten Stickstoffmonoxid-sensitiven Proteinen, der kleinen Untereinheit von *Glutamate synthase*, getestet. *In vivo* und *in vitro* Aktivitätsstudien zeigten, dass es zu einer schnellen Inaktivierung von *Glutamate synthase* nach NO-Behandlung kommt, das Protein aber resistent gegenüber anderen Oxidationsmitteln ist. Diese Resultate implizieren, dass reaktive Sauerstoff- und Stickstoffspezies unterschiedliche physiologische Vorgänge in Bakterien beeinflussen.

6 Zusammenfassung

Das zweite Ziel meiner Arbeit war es, redox-sensitive Proteine in *Saccharomyces. cerevisiae* (*S. cerevisiae*) zu identifizieren und deren Redox-Zustand als *in vivo* Read-Out zu verwenden, um Beginn, Ausmaß und die Rolle von oxidativen Stress während des Alterungsprozess eukaryotischer Zellen zu analysieren. Zunächst bestimmte ich in exponentiell wachsenden *S. cerevisiae* Zellen mit Hilfe von OxICAT, einer hochsensiblen quantitativen Methode, die Thiol-Trapping mit Massen-spektrometrie verbindet, den exakten *in vivo* Thiol-Status von fast 300 Proteinen. Diese Proteine, die sich sowohl im Cytosol als auch in verschiedenen Organellen (Mitochondrien, Zellkerne oder Vakuolen) befinden, lassen sich in vier Gruppen einteilen: 1) Proteine, deren Cysteinreste resistent gegen Oxidation sind; 2) Proteine, in denen Cysteinmodifikationen, vermutlich Disulfidbrücken, strukturell und funktionell wichtige Aufgaben übernehmen; 3) Proteine mit äußerst oxidationsempfindlichen Cysteinen, die aufgrund ihrer Sensibilität gegenüber geringen Mengen von reaktiven Sauerstoffspezies bereits eine gewisse Oxidation in exponentiell wachsenden Hefezellen aufweisen; 4) Proteine, die in exponentiell wachsenden Zellen reduziert sind, aber redox-sensitive Cysteinreste enthalten, die die katalytische Funktion der Proteine bei Vorhandensein von oxidativen Stress beeinflussen. Die Sensitivität dieser Proteine gegenüber oxidativen Stress wurde durch die kurzfristige Exposition subletaler Konzentrationen der Oxidationsmittel H_2O_2 oder Superoxid auf exponentiell wachsende Hefezellen nachgewiesen. Es wurde gezeigt, dass die wichtigsten zellulären Angriffspunkte von H_2O_2- und Superoxid-bedingtem Stress Proteine sind, die an Vorgängen der Translation, Glykolyse, des Citratzyklus und der Aminosäure-Biosynthese beteiligt sind. Diese Zielproteine zeigen, dass Zellen für die Bekämpfung von oxidativen Stress Metabolite und Energie schnell in Richtung des Pentosephosphatweges umleiten, um die fortwährende Produktion des Reduktionsmittels NADPH sicherzustellen. Die hier präsentierten Ergebnisse belegen, dass die quantitative Bestimmung des Oxidationsstatus von Proteinen eine wertvolle Methode ist, um katalytisch aktive und redox-sensitive Cysteinreste zu identifizieren. Zudem veranschaulicht diese Studie, wie Zellen reversible posttranslationale Thiol-Modifikationen nutzen, um eine Reaktion gegen die Auswirkung von oxidativen Stress zu koordinieren, die das Überleben der Zelle sichern soll.

Die OxICAT Technologie wurde dann verwendet, um das genaue Ausmaß und die Entstehung von oxidativen Stress in chronologisch alternden *S. cerevisiae* Zellen zu bestimmen. Für diese Bestimmung wurde der Oxidationsstatus von Proteinen in alternden Hefezellen als physiologischer Read-Out verwendet. Ich zeige, dass die zelluläre Redox-

Homöostase in chronologisch alternden Hefezellen global zusammenbricht, wobei es sich dabei um einen Prozess handelt, der dem Zelltod vorausgeht. Der Beginn dieses Zusammenbruchs scheint mit der Lebensdauer der Hefezellen zu korrelieren, da Kalorienrestriktion, welche durch eine Verringerung der Verfügbarkeit von Nährstoffen ohne Mangelernährung definiert ist, die Lebensdauer der Hefezellen erhöht und den Zusammenbruch des Redox-Gleichgewichts verzögert. Diese Ergebnisse deuten darauf hin, dass es sich bei der Aufrechterhaltung des Redox-Gleichgewichts um einen Teil der lebenserweiternden Maßnahmen von Kalorienrestriktion in Hefezellen handelt. Eine Gruppierungsanalyse aller Proteine, die in chronologisch alternden Hefezellen oxidativ modifiziert worden sind, ließ eine kleine Anzahl an Proteinen erkennen, deren oxidative Thiol-Modifikationen deutlich dem allgemeinen Redox-Zusammenbruch vorausgehen. Oxidation dieser Proteine führt vermutlich zu einem Rückgang ihrer Aktivität und könnte maßgeblich zum Verlust der Redox-Homöostase (z.B. im Fall von *Thioredoxin reductase*) oder des Zelltodes (z.B. im Fall von *Cell division control protein 48*) in chronologisch alternden Hefezellen beitragen. Diese Studien an alternden Hefezellen erweitern unser Verständnis, wie sich oxidativer Stress und Veränderungen in der Redox-Homöostase auf die Lebensdauer von Hefezellen auswirken. Zudem bestätigen die hier präsentierten Ergebnisse die Bedeutung von oxidativen Thiol-Modifikationen als eine der wichtigsten posttranslationalen Proteinmodifikationen in pro-und eukaryotischen Organismen.

1   Introduction

Abstract

For many years, oxidative thiol modifications in cytosolic proteins were largely disregarded as *in vitro* artifacts, and considered unlikely to play significant roles within the reducing environment of the cell. Recent developments in *in vivo* thiol trapping technology combined with *in vivo* mass spectrometric analysis have now provided convincing evidence that thiol-based redox switches are used as molecular tools in many proteins to regulate their activity in response to reactive oxygen and nitrogen species. Reversible oxidative thiol modifications have been found to modulate the function of proteins involved in many different pathways, starting from gene transcription, translation and protein folding, to metabolism, signal transduction, and ultimately apoptosis. This introduction will focus on three well-characterized eukaryotic proteins that use thiol-based redox switches to influence gene transcription, metabolism and signal transduction. The transcription factor Yap1 is a good illustration of how oxidative modifications affect the function of a protein without changing its activity. We use glyceraldehyde-3-phosphate dehydrogenase to demonstrate how thiol modification of an active site cysteine re-routes metabolic pathways and converts a metabolic enzyme into a pro-apoptotic factor. Finally, we introduce the redox-sensitive protein tyrosine phosphatase PTP1B to illustrate that reversibility is one of the fundamental aspects of redox-regulation.

1.1 Reactive oxygen species and oxidative stress

Reactive oxygen species (ROS) include a broad variety of different chemical oxidants including hydrogen peroxide (H_2O_2), superoxide anion ($\cdot O_2^-$), singlet oxygen ($^1O_2$), and hydroxyl radicals ($\cdot OH$). Inside the cells, reactive oxygen species are constantly generated as side products of neutrophil-mediated phagocytosis [2], cellular respiration [3], and various NADPH oxidases [4]. This ROS production appears not to be just a harmful byproduct of aerobic metabolism, but seems to involve a tightly regulated process [5]. It allows low levels of ROS to serve as important signaling molecules in processes such as gene transcription, protein tyrosine de- and phosphorylation, stress protection, apoptosis, and metabolism [5-8]. Similarly, reactive nitrogen species (RNS), which include the free radical nitric oxide ($\cdot NO$), peroxynitrite ($ONOO^-$) and nitrogen dioxide (NO_2), also play significant roles as redox-active molecules [9]. In the presence of oxygen, endogenously produced $\cdot NO$ can rapidly lead to S-nitrosation of protein thiols (SNO) and glutathione (GSNO), presumably via the formation of other reactive nitrogen species, such as N_2O_3, NO_2 and thiyl radicals [10]. Over 200 different proteins have been found to be targeted by S-nitrosation, including metabolic enzymes, phosphatases, transcription factors, and others [11].

To maintain redox homeostasis under conditions where ROS and RNS concentrations begin to increase (see below), cells can draw on a wide arsenal of enzymatic and non-enzymatic defense and repair strategies. For rapid ROS detoxification and scavenging, most cells use a combination of antioxidant enzymes with very high catalytic activity, such as superoxide dismutase (Sod), catalase (Cat), and glutathione peroxidase (Gpx), or of high cellular abundance, such as peroxiredoxins (>0.5% of total cell protein) [12]. The original oxidation status of cytosolic protein thiols is maintained and rapidly restored by the action of two redox-balancing systems, thioredoxin (Trx) and glutaredoxin (Grx), which draw their reducing power ultimately from NADPH [13]. The overall reducing environment of the cytosol of −220 to −260 mV, which represents the cellular standard redox potential, is preserved by the equilibrium between the small tripeptide glutathione (GSH) and its oxidized form (GSSG) [14,15]. The GSH/GSSG couple constitutes the redox buffer of most eukaryotic and prokaryotic cells. The low reactivity of the GSH thiol at physiological conditions (pKa of 9.4) [16] is counterbalanced by its high intracellular concentration (~1-10 mM) and rapid enzymatic regeneration by glutathione reductase and NADPH [17].

Once ROS and RNS concentrations exceed the antioxidant capacity of the cell, oxidative damage to DNA, lipids, and proteins can occur [8,18,19]. This puts organisms into a state defined as oxidative stress [20]. It triggers a highly conserved response in both pro- and eukaryotic cells [21]. The response includes increased expression of oxidant scavengers, chaperones, proteases, and DNA repair enzymes to mitigate and reverse oxidative damage on proteins and DNA. Severe oxidative damage, which will eventually lead to apoptosis and cell death, has been shown to accompany many pathological conditions, such as cancer, neurodegenerative diseases, and diabetes [22,23]. Moreover, it is the damaging effect of increasing oxidant concentrations that is thought to be the underlying culprit of eukaryotic aging [5].

1.2   Cysteine thiols — Central components of redox-sensitive nanoswitches

Cysteine residues play important roles in the biochemistry of many proteins due to the unique chemical properties of their side chain. The high reactivity, redox properties, and ability of thiol groups to coordinate metal ions designate cysteines as the amino acids of choice to form key catalytic components of enzymes [24]. Over the past ten years, however, a new role for cysteine thiols emerged. They form the central building block of thiol-based redox switches in redox-regulated proteins. Redox-sensitive cysteines undergo reversible thiol modifications in response to reactive oxygen or nitrogen species, thereby modulating protein function, activity, or localization. They play important roles in many signal transduction cascades (e.g., PTEN and PTP1B) [25,26], gene transcription (e.g., p53) [27], as well as in the immediate defense against oxidative stress conditions (e.g., Hsp33, Hsf-1) [28,29].

What makes cysteine residues particularly redox-sensitive and proteins potentially redox-regulated? Regulatory thiol modifications involve one or more cysteines, whose reactivity is largely determined by the cysteine's structural environment and its pKa value. Most cytoplasmic protein thiols have pKa values of greater than 8.0, which render the thiol groups predominantly protonated and largely non-reactive at intracellular pH [30]. Thiol groups of redox-sensitive cysteines, on the other hand, have characteristically much lower pKa values, ranging from as low as ~3.5 in thiol transferase to ~5.1 - 5.6 in protein tyrosine phosphatases. Under physiological pH conditions, these thiols are therefore present as deprotonated, highly reactive thiolate anions (RS$^-$) [31,32]. The low pKa values of redox-sensitive cysteines arise primarily from stabilizing charge-charge interactions between the thiolate anion and neighboring positively charged or aromatic side chains [33,34].

Thiolate anions, in contrast to their protonated counterparts, are highly susceptible to oxidation by ROS and RNS and can undergo a diverse spectrum of oxidative modifications [35]. These include sulfenic (R-SOH), sulfinic (R-SO$_2$H), and sulfonic (R-SO$_3$H) acids, disulfide bonds (Pr-SS-Pr), or nitrosothiols (SNO) [8,36,37] (Fig. 1.1). Cysteine sulfenic acids and their deprotonated cysteine-sulfenates are remarkably reactive and versatile oxidation products, which are frequently formed first upon reaction of protein thiols with H$_2$O$_2$ [38]. Due to their high reactivity, sulfenic acids are often considered as metastable intermediates that undergo further reactions to form stable modifications, such as disulfides with other protein thiols or S-glutathionylation with the small redox buffer component glutathione [8,36-38]. Importantly, most oxidative thiol modifications are fully reversible *in vivo* and utilize dedicated oxidoreductases, such as the thioredoxin or glutaredoxin system, to quickly restore the original redox state upon the cell's return to non-stress conditions [39-41] (Fig. 1.1). It appears that it is the reaction rate with these dedicated oxidoreductases that often determines the life span of oxidized proteins and supports their accumulation even in an overall reducing environment [42,43].

Extended and/or extensive exposure of proteins to oxidants can lead to the "overoxidation" of cysteine residues to form sulfinic and sulfonic acids [44]. These oxidative modifications are considered largely irreversible *in vivo*, although a small number of protein-specific sulfinic acid reductases (e.g., sulfiredoxin, sestrin) have recently been identified [45,46]. Moreover, strong oxidative stress conditions often result in excessive disulfide bonding, protein misfolding, aggregation, and degradation, and will eventually lead to cell death [47,48].

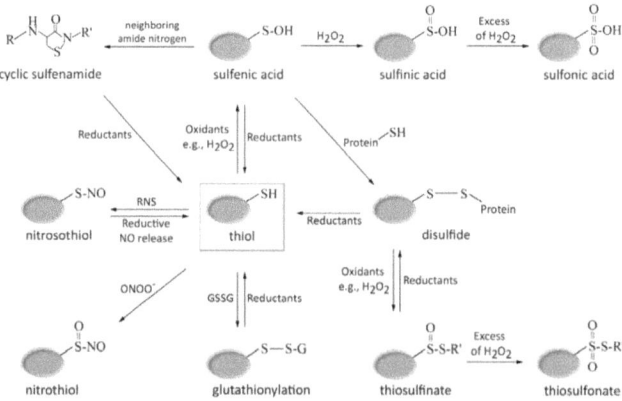

Figure 1.1. Oxidative thiol modifications. *Oxidation of cysteine thiol groups by H_2O_2 leads to sulfenic acid (R-SOH) formation. Sulfenic acids are either stabilized by nearby charges or react with neighboring thiols or proximal nitrogen to form disulfide bonds (R'-S-S-R') or sulfenamide bonds (R'-S-NH-R"), respectively. In the presence of high H_2O_2 concentrations, overoxidation to sulfinic (R-SO$_2$H) or sulfonic acid (R-SO$_3$H) occurs. Although a few protein-specific sulfinic acid reductases have been identified, overoxidation is still considered to be largely irreversible in vivo. Alternatively, reaction of thiolate anions (RS$^-$) with oxidized cysteines of other proteins or low molecular weight thiols such as glutathione (GSSG) leads to mixed disulfide bond formation (R'-S-S-R") or S-glutathionylation (R-S-SG), respectively. Overoxidation of disulfide bonds in the presence of strong oxidants can cause thiosulfinate (R'-SO-S-R") or irreversible thiosulfonate (R'-SO$_2$-S-R") formation. Most oxidative thiol modifications are reversed by members of the glutaredoxin (Grx) system and thioredoxin (Trx) system (reductants), which draw their reducing power from cellular NADPH. Exposure of thiolate anions to reactive nitrogen species causes S-nitrosothiol formation, whereas treatment with peroxynitrite yields S-nitrothiol formation. The exact mechanism by which individual RNS cause oxidative thiol modifications in vivo is still under investigation.*

The type and extent of oxidative modifications in redox-regulated proteins depends on the specific type of oxidant [49], its absolute level and persistency, the subcellular location of oxidant production and its distance to the target protein [5,6]. Even small changes in the basal level of intracellular ROS and RNS can cause oxidative modifications in proteins that are specifically sensitive to these oxidants. Oxidative thiol modifications most often lead to changes in the structure of proteins and cause either their activation (e.g., OxyR, Hsp33) [29,50] or inactivation (e.g., PTEN, GapDH) [26,51]. With the help of innovative *in vivo* and *in vitro* thiol trapping techniques (reviewed in: [52,53]) and global redox proteomic methods (reviewed in: [54]), the number of redox-regulated proteins that have been identified is steadily increasing [37,55,56]. Redox-regulated proteins turn out to be involved in nearly every physiological process, including metabolism, signaling, cell growth, gene expression, transcription factor activation, differentiation, senescence, and apoptosis [5,7,8,36]. Interestingly, more often than not, redox-regulated proteins appear to be central components of these pathways, suggesting that protein redox regulation is the cell's fundamental strategy to fine-tune cellular processes in response to ROS and RNS. Here, we focus on a subset of proteins with thiol-based redox switches, whose redox regulation influences distinct processes in eukaryotic cells: transcription, metabolism, and signal transduction.

1.3   Changing gene transcription to protect cells against oxidative stress

The ability to rapidly adapt gene expression to environmental changes is crucial for the growth and survival of every organism. Over the past few years, an increasing number of transcription factors have been shown to be directly regulated by accumulating ROS [55]. These include the activator protein-1 (AP-1), NF-κB, Myb, p53, USF (upstream stimulatory factor), NF-1 (nuclear factor 1) and others [57-59]. Redox regulation of these proteins is mediated by the modification of one or more cysteine residues, which causes either the activation (e.g., AP-1, NF-κB) or inactivation (e.g., USF) of the respective transcription factor [57,59].

We focus here primarily on AP-1-like transcription factors, which are well-characterized leucine-zipper (bZIP) DNA-binding proteins, which are implicated in the regulation of multiple cellular processes, including proliferation, differentiation, stress response and apoptosis [60-62]. In *Saccharomyces cerevisiae*, the AP-1-like transcription factor Yap1 plays an important role in the cellular response to oxidative stress and various xenobiotics [63-65]. The Yap1 regulon includes as many as 70 genes encoding most of yeast's antioxidant enzymes and components of the cellular thiol-reducing pathway, including Trx, Grx, Gpx, Sod, and Cat [63,66,67].

1.4   Yap1 — Changing the redox state means switching homes

The activation of Yap1 upon peroxide stress (i.e., H_2O_2) is triggered by highly conserved cysteine residues, which are reduced under non-stress conditions and rapidly form intramolecular disulfide bonds upon exposure to oxidative stress conditions [68,69]. The redox-active cysteines in Yap1 are clustered in two cysteine-rich domains (CRD) (Fig. 1.2). Each CRD contains three cysteines (Cys303, Cys310, and Cys315) positioned in the n-

CRD at the center of the protein, and three cysteines (Cys598, Cys620, and Cys629) located in the c-CRD at the C-terminus of Yap1. Embedded in the c-CRD is also a leucine-rich nuclear export signal (NES), which is directly responsible for Yap1's subcellular distribution and indirectly responsible for its functional regulation [70,71].

Yap1's function as a transcription factor is controlled by redox-regulated changes in the accessibility of NES [70,72]. In non-stressed yeast cells, the NES is exposed and Yap1 shuttles between the nucleus and the cytosol. Upon migration into the nucleus, Yap1 is rapidly exported in a process that is mediated by binding to the nuclear export receptor Crm1 (also called Xpo1) [71,72]. The outcome is a relatively low nuclear level of Yap1 and therefore a low constitutive activity in unstressed yeast cells. Exposure of yeast cells to H_2O_2, however, leads to rapid redox-mediated conformational changes in Yap1, which disrupt the interaction with Crm1. This, in turn, leads to the nuclear accumulation of Yap1 and results in significantly increased antioxidant gene transcription [68,70].

1.4.1 Yap1 — Member of a two-component redox relay

Over the past few years it became evident that Yap1 is not directly oxidized by H_2O_2 but functions in a two-component system with its partner protein glutathione peroxidase Gpx3 (also named Orp1), a protein originally identified as a hydroperoxide scavenger [66]. Upon exposure to H_2O_2, the catalytic site cysteine Cys36 of Gpx3 attacks the O-O bond of the substrate, thereby forming a sulfenic acid (Cys36-SOH) at its active site cysteine [32] (Fig. 1.2). Cys36-SOH subsequently reacts with Cys598 of Yap1's c-CRD to form a mixed Yap1-Gpx3 disulfide intermediate. Rearrangement of this disulfide by intramolecular thiol-disulfide exchange with Cys303 of Yap1 then leads to the formation of the first inter-domain disulfide bond between Cys303-Cys598 in Yap1 and the regeneration of reduced Gpx3 [32,73]. Triggered by this inter-domain disulfide bond formation, Yap1 undergoes further conformational changes that apparently mask the NES and disrupt the Yap1-Crm1 interaction [68,74] (Fig. 1.2) Subsequently, additional disulfide bonds between cysteines of n-CRD and c-CRD of Yap1 form (see below) presumably in a similar, Gpx3-mediated process [6]. Upon return to non-stress conditions, the disulfide bonds are reduced by the thioredoxin (Trx) system, leading to the reversal of the structural rearrangements and to the re-exposure of NES [68]. Thus, Crm1-mediated nuclear export of Yap1 can resume and Yap1's transcriptional activity returns to its low pre-stress levels [69].

Chapter 1: Introduction

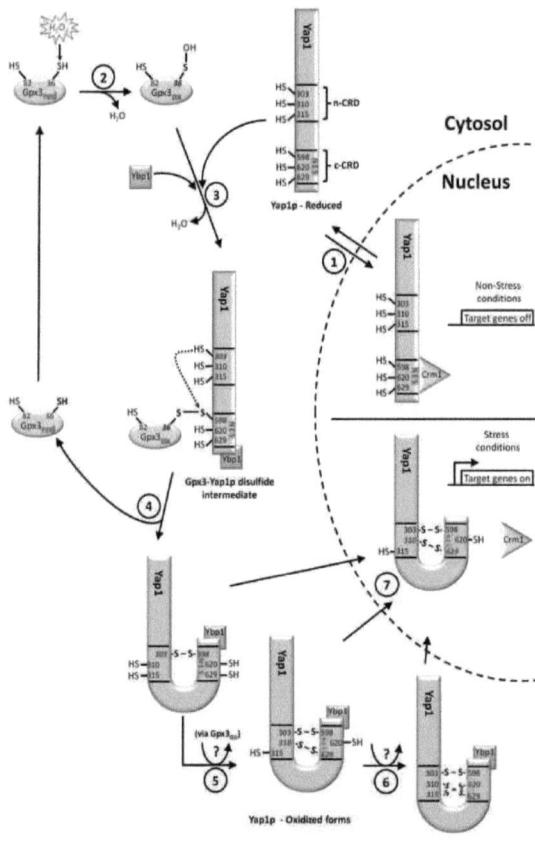

Figure 1.2. Model of the Yap1-Gpx3 redox relay. *Under non-stress conditions, the cysteines of yeast Yap1, which are clustered in two cysteine-rich domains (n-CRD, c-CRD), are reduced.* **(1)** *The nuclear export sequence (NES) of Yap1 is accessible for interaction with Crm1, and Yap1 shuttles between the cytosol and the nucleus. It prevents Yap1's nuclear accumulation and guarantees low constitutive activity under non-stress conditions.* **(2)** *Upon exposure of yeast cells to oxidative stress conditions (i.e., H_2O_2) the active site cysteine (Cys36) of glutathione peroxidase Gpx3 attacks the O-O bond of H_2O_2, thereby forming a sulfenic acid at the active site cysteine. At the same time, Ybp1 binds to Yap1.* **(3)** *In the next step, the active site sulfenic acid of Gpx reacts with Cys598 of Yap1, causing the formation of a Gpx3-Yap1 disulfide inter-mediate.* **(4)** *A thiolate anion formed at Cys303 of Yap1 attacks the intermolecular disulfide bond, thereby causing the formation of the inter-domain Cys303-Cys598 disulfide bond and the recovery of reduced Gpx3.* **(5)** *Additional inter-domain disulfide bonds form in a process likely also involving Gpx3. While formation of the Cys310-Cys629 appears to increase transcriptional activation,* **(6)** *formation of Cys315-Cys620 might prolong the H_2O_2-mediated signal. The exact role of Ybp1 has yet to be determined, but it appears that Ybp1 mediates the redox signal from Gpx3 to Yap1.* **(7)** *Disulfide bond formation appears to lead to conformational changes in Yap1, which mask the NES and prevent nuclear export. Yap1 accumulates in the nucleus and activates the oxidative stress response. For reasons of simplicity, only one oxidized form of Yap1 is shown in the nucleus.*

A similar relay system has also been described for the redox-regulation of the mammalian AP-1 transcription factor that is formed upon dimerization of c-Fos and c-Jun. In contrast to Yap1, however, which requires the sulfenic acid in Gpx3 for its oxidative activation, mammalian AP-1 uses the reduced cysteines of the redox factor Ref-1 for its reductive activation [75,76]. In the c-Fos/c-Jun AP-1 transcription factor, a single conserved cysteine residue in the DNA-binding domains of each of the two individual proteins appears to form the molecular redox switch [77]. *In-vitro* treatment of AP-1 with alkylating agents, such as N-ethylmaleimide (NEM) has been shown to cause a decrease in transcriptional activity while exposure to reductants, such as DTT, or mutation of the cysteines results in enhanced DNA binding [77,78]. Once oxidized and inactivated, AP-1 cannot be directly reduced and activated by Trx [77]. Activation of AP-1 requires Ref-1, which appears to draw its electrons from Trx and ultimately from NADPH [79].

1.4.2 Redox-mediated fine-tuning of Yap1's functional activity

In vitro studies demonstrated that Yap1's activation by H_2O_2 involves a multi-step formation of three inter-domain disulfide bonds. This finding led to a model suggesting that all six cysteine residues in Yap1 are important for optimal H_2O_2 sensing and signal transduction [74]. Transcriptional activity of Yap1 appears to be directly determined by the balance between Gpx3-mediated disulfide bond formation in Yap1, which is regulated by intracellular H_2O_2-levels, and thioredoxin-mediated disulfide bond reduction, which is regulated by intracellular NADPH levels [73], suggesting that the cellular ratio of H_2O_2 to NADPH ultimately determines the number of disulfide bonds that form and the level of Yap1 activity [74].

In vitro activation generates at least three oxidation forms of Yap1, which differ in the number of disulfide bonds [74]. As mentioned above, formation of the Cys303-Cys598 disulfide bond seems to initiate the activation of Yap1 by directly causing conformational changes that bury the NES. Formation of the second inter-domain disulfide bond between Cys310 and Cys629 appears to further increase Yap1's transcriptional activity. Both inter-domain disulfide bonds seem to be required for the recruitment of Rox3p to the Trx2 promoter, a prerequisite for maximum Yap1-mediated Trx2 expression [80]. Moreover, Yap1 mutant strains carrying substitutions at either Cys310 or Cys629 show a mixture of oxidized and reduced Yap1 species upon H_2O_2 stress *in vivo*. This result suggests that the formation of the Cys310-Cys629 bond stabilizes the Cys303-Cys598 disulfide and prevents its premature reduction [69]. The arrangement of the third disulfide bond between Cys315 and Cys620 appears to be important for maintaining Yap1 in the nucleus and upholding its activity in the later stages of the H_2O_2 response [74].

Noteworthy at this point is the observation that a third protein appears to be involved in peroxide-induced Yap1 oxidation. *In vivo* studies showed that the protein Ybp1 forms an

H_2O_2-induced complex with the c-CRD-containing region of Yap1, thereby stimulating nuclear accumulation and activity of the transcription factor [81]. Although Ybp1 has several cysteine residues, this interaction appears to be based on a non-redox function of Ybp1 [6]. Because Δ*ybp1*/Δ*gpx3* double mutants show very similar effects on the regulation of Yap1's activity as the individual single mutants, a model was proposed in which Ybp1 and Gpx3 act in the same pathway [81] (Fig. 1.2). The H_2O_2 stress-induced stimulation of the Yap1-Ybp1 complex formation, together with the cytoplasmic localization of Ybp1, suggests that the molecular function of Ybp1 might be important for disulfide bond formation or signal transduction between Gpx3 and Yap1 [81,82]. Alternatively, Ybp1 might function as mediator in the formation of the Gpx3-Yap1 disulfide intermediate or in preventing Gpx's sulfenic acid intermediate from reacting with its own second cysteine, which would short-cut the oxidation relay.

1.4.3 Activation of Yap1 — More than one way to get the response

In contrast to the activation of Yap1 in response to ROS, such as peroxide, which depends on Gpx3 and/or Ybp1, activation of Yap1 in response to diamide, electrophiles, and divalent heavy metal cations appears to function independently of Gpx3 and does not involve any inter-domain disulfide bonds [65,68]. In the presence of diamide, thiol modification of the three cysteines in c-CRD seems sufficient to disrupt the interaction between Yap1 and Crm1 and to promote the nuclear accumulation of Yap1 [83]. Diamide-mediated thiol modifications might either lead to the formation of an intra-domain disulfide bond between two of the three cysteines in c-CRD [83] or cause individual cysteine modification, such as the formation of sulfenylhydrazine [65,68,83]. Activation of Yap1 without disulfide bond formation was also observed in response to the superoxide generator and alkylating agent menadione and N-ethylmaleimide under anaerobic conditions as well as in a Δ*gpx3* strain [65]. Based on these findings, it is likely that other non-ROS inducers of Yap1, such as cadmium or selenate activate Yap1 by either binding to or chemically modifying individual C-terminal cysteines [65,68]. This type of activation is similar to the diethylmaleate-mediated activation of the Yap1-homologue Pap1 in *Saccharomyces pombe*, which is suggested to operate through covalent cysteine adduct formation [84].

1.4.4 Yap1 — A prototype for emerging concepts in redox regulation?

Yap1's mechanism of activation combines a number of features that have become increasingly common among redox-regulated proteins. One concept that emerges is that redox-regulated disulfide bond formation causes extensive conformational changes in the affected proteins, which either lead to the folding of previously unfolded regions (e.g., Yap1, Pap1) [69,84] or to the unfolding of previously folded regions (e.g., Hsp33, RsrA) [29,85]. Whether disulfide bond formation mediates folding or unfolding of the protein depends largely on the individual protein, its function, and the mechanism of regulation. Interestingly, it appears not to correlate with the final activity of the oxidized protein. Disulfide bond formation in Yap1, for instance, converts a poorly folded region into a compactly folded protein domain in a process that effectively buries the NES and activates the transcriptional response [69]. In contrast, ROS-mediated disulfide bond formation in the redox-regulated chaperone Hsp33 causes large portions of the protein to adopt a natively unfolded protein conformation [29]. This unfolding is crucial for the exposure of a substrate-binding site in Hsp33, which is capable of binding proteins that undergo stress-induced unfolding and protects bacteria against oxidative damage [29]. In the anti-sigma factor RsrA, on the other hand, disulfide-mediated unfolding abolishes its binding to the sigma factor, which is released and induces antioxidant gene transcription in *Streptomyces coelicolor* [85]. Extensive biophysical studies on other redox-regulated proteins will reveal how many proteins use this rapid and fully reversible mechanism to translate changes in their redox state into functional changes.

A second mechanistic aspect that might turn into a new paradigm is that thiol-based regulatory switches such as found in Yap1 or the related Pap1p do not function as primary ROS sensors. Instead, both redox-regulated transcription factors use peroxidases as ROS sensors, which relay the redox signal from their active site cysteine to the acceptor protein and trigger disulfide bond formation [86]. It is conceivable that other thiol peroxidases serve as central relay stations for yet to be identified redox-regulated proteins *in vivo*.

1.4.5 Redox theme with variations — The Nrf2-Keap1 connection

In multicellular organisms, protection against ROS, nitric oxide, heavy metals, and electrophilic compounds is primarily mediated by the redox-sensitive Nrf2-Keap1 signaling pathway [87,88]. Nrf2, the nuclear factor erythroid-2 related factor, functions as master transcription factor. It forms heterodimers with other transcription factors such as c-Jun,

Jun B, Jun D, ATF3, and ATF-4, thereby regulating the transcription of more than 200 eukaryotic genes. The gene products are involved in the cellular protection against oxidative stress, in the detoxification of electrophiles, and in proteasomal activity [88].

Under non-stress conditions, Nrf2 forms a tight complex with the redox-sensitive stress receptor protein Keap1 (Kelch-like ECH-associated protein-1). Keap1 functions as a negative regulator of Nrf2 by retaining the transcription factor in the cytoplasm and targeting it for ubiquitin-mediated proteasomal degradation [89]. Keap1 facilitates Nrf2's proteasomal degradation by acting as a substrate adaptor protein for Cul-containing E3 ubiquitin ligases, which catalyze the ubiquitination and subsequent proteasomal degradation of Nrf2 (reviewed in [90]). Exposure of cells to ROS or other inducers, however, promotes the escape of Nrf2 and its subsequent translocation into the nucleus, where it activates the antioxidant response element (ARE) [91].

Stress sensing is mediated by the cysteine-rich protein Keap1. At least three (Cys151, Cys273, Cys288) of the 25 to 27 cysteines present in mammalian Keap1 appear to play critical roles. Interestingly, any type of thiol modifications at these three cysteine residues seem to be sufficient to induce conformational rearrangements in Keap1, which abrogate its binding to Nrf2 and activate the oxidative stress response [88]. Recently, Keap1 was identified as a metal-binding protein, which binds zinc in a 1:1 stochiometry [87]. Mutant studies suggested that two of the three redox-sensitive cysteines (Cys273 and Cys288) are directly involved in zinc coordination under reducing non-stress conditions [87]. Exposure to ROS was found to induce disulfide bond formation and lead to zinc release and conformational rearrangements in Keap1, suggesting that Keap1 senses ROS via a redox-sensitive zinc center. This model is very reminiscent of the redox-regulated anti-sigma factor RsrA [85] and the bacterial chaperone Hsp33 [29]. Both proteins use redox-sensitive zinc-centers as ROS-sensors and undergo dramatic conformational changes upon disulfide bond formation and zinc release. The additional layers of regulation that are involved in the Nrf2-Keap1 signaling pathway have been expertly reviewed in the past [90], and their discussion goes beyond the scope of this introduction.

1.5 GapDH — Putting cellular metabolism under redox-control

A variety of key metabolic enzymes in eukaryotic cells have been identified to contain redox-sensitive active site cysteine residues, which become oxidatively modified upon exposure to ROS and/or RNS. These redox-regulated metabolic enzymes include among others carbonic anhydrase, creatine kinase (CK), and glyeraldehyde-3-phosphate dehydrogenase (GapDH) [92-94]. Here, we focus on GapDH, which has a central function in glycolysis in both prokaryotic and eukaryotic cells. Glycolysis is the almost universal central pathway of glucose catabolism and represents the main source of metabolic energy for many cells and organisms. GapDH catalyzes the reversible oxidative phosphorylation of glyeraldehyde-3-phosphate (GAP) to 1,3-bisphosphoglycerate in the presence of nicotinamide adenine dinucleotide and inorganic phosphate. It uses a highly conserved and reactive active site cysteine (Cys149 or Cys150, respectively) for the initial nucleophilic attack on the aldehyde of GAP [95]. This active site cysteine has been shown to be highly sensitive to oxidative modifications by a variety of different reactive oxygen and nitrogen species. In addition, GapDH appears to play an important role in a plethora of other processes that are unrelated to its glycolytic function [96]. These include transcription [97], regulation of Ca^{2+}-homeostasis [98], membrane fusion, microtubule bundling, phosphotransferase activity, nuclear RNA export, and DNA replication and repair [96,99]. Recent studies have furthermore shown an active role of GapDH in apoptosis [100] and an involvement in Huntington's [101], Parkinson's [102], and Alzheimer's disease [103]. The pro-apoptotic role of GapDH in these diseases seems to be dependent upon its nuclear accumulation and its formation of insoluble aggregates [48,104].

GapDH functions as a homotetramer. Each subunit contains one active site cysteine and a second, somewhat less conserved cysteine three amino acids apart. Upon exposure to oxidative stress conditions, the active site thiolate in GapDH was shown to rapidly become the target of both reversible and irreversible thiol modifications, whose nature largely depends on the type and concentration of oxidants. In the presence of H_2O_2, for instance, GapDH's active site cysteine was found to be oxidized to either sulfenic, sulfinic, sulfonic acid, or engaged in an intramolecular disulfide bond with the neighboring Cys154 [49]. Incubation with HNO (nitroxyl anion) or the NO-donor DEA/NO (diethylamine NONOate), on the other hand, was shown to cause S-nitrosation [94,100], S-glutathionylation [40,51,94], carbonylation [105], and ADP-ribosylation [106]. All these modifications cause the reversible or irreversible inhibition of GapDH enzyme activity and lead to the attenuation of glycolysis *in vivo*.

1.5.1 Re-Routing the metabolic flux to protect cells against oxidative damage

Oxidative stress-induced inhibition of GapDH as a central component of glycolysis leads to the rapid re-routing of glucose-6-phosphate into the pentose phosphate pathway (PPP). This serves as immediate response to protect against oxidative insults [107,108] (Fig. 1.3). The PPP is one of the key routes for nicotinamide adenine dinucleotide phosphate (NADPH) production in the eukaryotic cytoplasm. The product NADPH serves as electron donor for both glutathione reductase (Grx) and thioredoxin reductase (Trr), and is therefore essential to restore and maintain the reducing environment of the cytosol [109]. Oxidative inactivation of triosephosphate isomerase (Tpi) causes a similar re-direction of the metabolic flux from glycolysis to PPP. Tpi, which directly precedes GapDH in glycolysis, catalyzes the isomerization of dihydroxyacetone phosphate to GAP. Yeast cells with reduced Tpi activity are highly resistant to treatment with the oxidant diamide [110], supporting the model that re-routing the glucose flux maintains the cytoplasmic NADPH/NADP$^+$ equilibrium and counteracts oxidative stress. When cells return to non-stress conditions, reduced glutathione and thioredoxin [42] rapidly restore GapDH activity and therefore the fine-tuned metabolic flux of glucose.

These findings show that ROS-mediated changes in the activity of glycolytic enzymes trigger important and immediate response mechanisms that control the redox state of the cell and protect cells against increased oxidative damage. These observations also form the basis for discussions that inhibitors of glycolysis such as 2-deoxy-D-glucose [111] might serve as potential therapeutics for oxidative stress-related neuronal disorders such as Alzheimer's and Parkinson's disease [107].

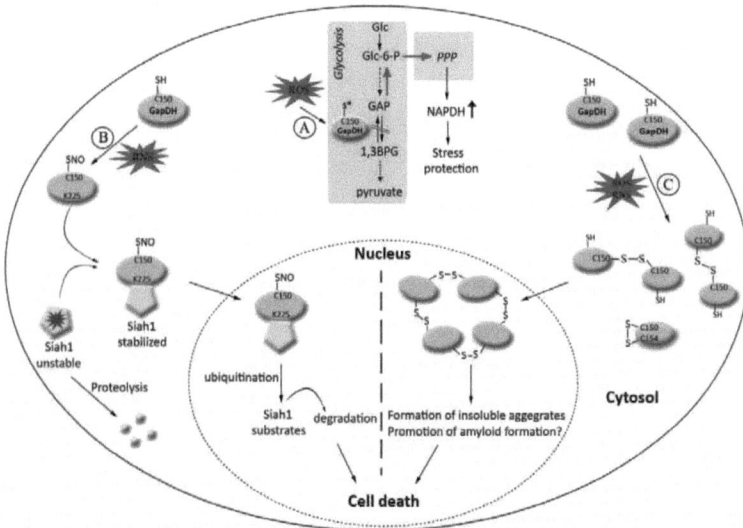

Figure 1.3. Oxidative thiol modifications of GapDH — From cytoprotection to cytotoxicity.
(A) ROS-mediated inhibition of GapDH re-routes metabolic flux as immediate defense against an oxidative insult. Oxidative stress-induced inhibition of GapDH as a central component of glycolysis leads to the rapid re-routing of glucose-6-phosphate into the pentose phosphate pathway (PPP), one of the key routes for NADPH production in the eukaryotic cytoplasm. NADPH serves as electron donor for glutathione reductase and thioredoxin reductase, and is therefore essential to restore and maintain the reducing environment of the cytosol. *(B)* Model of GapDH/Siah1-signaling cascade-dependent cell death. Exposure of GapDH to RNS causes S-nitrosation of Cys150 in GapDH. S-nitrosation stimulates the interaction between the E3-ubiquitin ligase Siah1 and GapDH. Siah1, which possesses a nuclear localization signal (NLS), escorts GapDH into the nucleus where the association between GapDH and Siah1 appears to stabilize the E3-ligase. This stabilization, in turn, increases the rate of degradation for a set of different target proteins, and ultimately triggers cell death. *(C)* Model of apoptosis triggered by ROS-mediated GapDH aggregation. Exposure of GapDH to excess ROS leads to the initial disulfide-bond formation between either Cys150 and Cys154 (intramolecular disulfide bond) or between two active site cysteines Cys150. Latter leads to extensive conformational changes and exposes additional cysteines (e.g., Cys281) that undergo further intermolecular disulfide bond formation. This results in the accumulation of insoluble aggregates composed of multiple GapDH subunits in the cytosol and the nucleus. The aggregates are either intrinsically cytotoxic or interact with other aggregation-prone proteins (e.g., Aβ, tau, α-synuclein) to cause cytotoxicity. For reasons of simplicity, only one subunit of the GapDH tetramer is shown.

1.5.2 ROS-mediated GapDH aggregation and its role in apoptosis

Both nitrosative and oxidative stress conditions have been shown to lead to intermolecular disulfide bond formation between GapDH subunits, causing extensive GapDH oligomerization both *in vitro* and *in vivo* [48,112] (Fig. 1.3). Interestingly, these aggregates were very similar in their characteristic features (e.g., Congo red staining) to amyloid-like fibrils previously reported for α-synuclein [113] and β-amyloid [114]. The ability of thiol reductants such as dithiothreitol (DTT) to dissolve these aggregates was dependent on the extent of oxidation, supporting the model that intermolecular disulfide bonds are involved in aggregate formation. Two cysteines appear to play the major role in this oxidative stress-induced aggregation process, the active site cysteine (Cys149 in rabbit GapDH, Cys150 in yeast GapDH) and, to a lesser extent, Cys281 [48,112]. The precise mechanism by which these large disulfide-linked GapDH oligomers form has yet to be determined. Given that the active site cysteine is located at the bottom of GapDH's catalytic site, the distance between these cysteines on the individual subunits is too far to allow inter-subunit disulfide bond formation to occur [112]. It is conceivable, however that initial thiol modification of Cys150 causes extensive conformational changes that lead to the exposure of additional cysteines such as Cys281 and allow them to undergo intermolecular disulfide bond formation [112].

Importantly, *in vivo* studies demonstrated that the degree of GapDH aggregation correlates well with the rate of oxidative stress-induced cell death [112]. These results suggested that insoluble GapDH aggregates are either cytotoxic per se, or promote apoptosis indirectly by interacting with other intracellular aggregate-prone proteins, such as tau or α-synuclein [115,116] (Fig. 1.3). This mechanism appears to contradict an earlier model proposed by Hara *et al.* in which NO-induced alterations of the GapDH/Siah1 signaling cascade were thought to cause the GapDH-dependent cell death under nitrosative stress conditions [100] (Fig. 1.3). According to this model, S-nitrosation of Cys150 in GapDH of HEK293 cells stimulates the interaction between the ATP-dependent E3-ubiquitin ligase Siah1, and Lys225 in GapDH [100,117]. Siah1, which possesses a nuclear localization signal (NLS), escorts GapDH into the nucleus where the association between GapDH and Siah1 appears to stabilize the E3-ligase [100]. Stabilization of Siah1, in turn, increases the rate of degradation for a set of different target proteins and ultimately triggers apoptosis [118] (Fig. 1.3). As so often in biology, it is more than likely that both mechanisms are linked to oxidative stress-induced cell death. More research is required to identify whether the two mechanisms work synergistically or independently, and to elucidate whether oxidant specificity of GapDH's active site cysteine determines the apoptotic route that cells take [100]. In either case, these studies demonstrate that GapDH, a protein best known for its central role in glycolysis, probably plays an equally important role as a redox-regulated pro-apoptotic protein [100,112].

1.5.3 Role of GapDH's redox-regulation in signaling

Nuclear translocation of GapDH mediated by oxidative modifications of its active site cysteine appears to correlate with the initiation of cell death and the physiology of a number of neurodegenerative diseases [48,100,102,104]. In addition, GapDH translocation might also play a role in intracellular signaling [99,100]. GapDH has long been shown to associate with DNA [119], and translocation into the nucleus might further stimulate this interaction. It is conceivable that oxidative stress-mediated nuclear translocation of GapDH serves as mediator for relaying extranuclear stress signals into the nucleus, which would translate changes in the cellular redox state directly into changes in transcriptional activity [120]. This is in line with data that show the direct involvement of metabolic enzymes or homologs such as GapDH or CtBP in transcription [97,121]. P38, a nuclear form of GapDH, was found to be a key component of OCA-S, a multicomponent Oct-1 co-activator complex, which is essential for S phase-dependent histone H2B transcription during DNA replication [97]. GapDH's direct interaction with the promoter-bound Oct-1 supports the binding of the OCA-S complex to Oct-1. Noteworthy, both the interaction of GapDH with Oct-1 as well as the transcriptional activity of OCA-S are altered by the NAD^+/NADH redox status [97]. This NAD^+/NADH influence might link the metabolic and/or redox state of the cell to histone transcription and DNA replication. However, the structural basis for NAD^+/NADH modulation and the mechanism by which GapDH supports gene transcription have yet to be determined [97].

Very recently the involvement of GapDH's redox regulation in a multi-step phosphorelay signaling in Schizo*saccharomyces pombe* was reported [122]. In phosphorelays, which are variations of the bacterial two-component signaling system, a histidine-containing phosphotransfer (HPt) protein mediates the phosphotransfer from a sensor kinase to a response regulator, which in turn activates a kinase cascade and eventually causes induction of the corresponding response genes [123]. In Schizo*saccharomyces pombe*, osmotic and oxidative stress conditions activate the mitogen-activated protein kinase (MAPK) Spc1 [124] (Fig.1.4). Activated Spc1 phosphorylates the Atf1 transcription factor, which induces several stress response genes, including *ctt1* encoding for cytoplasmic catalase [125]. The transmission of the stress signals to Spc1 in *S. pombe* occurs via a multi-step phosphorelay system consisting of the phosphotransfer protein Mpr1 and the response regulator Mcs4, which forms a complex with the stress-responsive Wis4/Win1 MAP kinase kinase kinase (MAPKKK) [126]. Morigasaki *et al.* showed that the Msc4-MAPKKK complex contains Tdh1, an isoform of GapDH [122]. Oxidative modification of Tdh1's active site cysteine Cys152 was found to result in enhanced binding between Tdh1 and Msc4, which appears to be essential for the interaction between Msc4 and Mpr1 and promotes the transmission of H_2O_2 stimuli to the downstream components of the MAP-kinase cascade (i.e., Wis1, Spc1) [126] (Fig. 1.4). Mutation of the active site cysteine (C152S) or deletion of the *tdh1* gene was demonstrated to cause a decrease in Msc4 and Mpr1 interaction, resulting in a defect in H_2O_2-signaling through the phosphorelay. These results reveal that peroxide-induced oxidative modification of the redox-sensitive cysteine in GapDH plays a key role in transmitting H_2O_2 stimuli to a response regulator, which is part of a multistep phosphorelay cascade, thereby modulating cellular signaling [122]. As GapDH is conserved in multiple eukaryotic organisms, GapDH might play an important role in other signaling relay systems in different species.

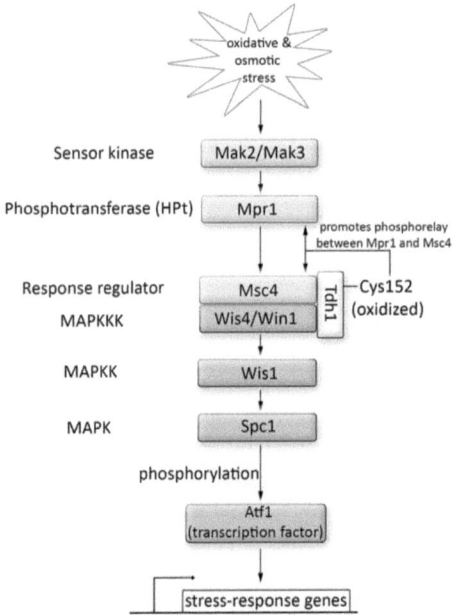

Figure 1.4. Role of GapDH's redox regulation in stress signaling. *In the fission yeast Schizosaccharomyces pombe, a multistep phosphorelay composed of the sensor kinase Mak2/Mak3, the histidine-containing phosphortrans-ferase (HPt) Mpr1 and the Msc4 response regulator transmit stress signals, such as H_2O_2, to the Spc1 MAPK cascade. This cascade is composed of the Wis4/Win1 MAPKKKs, the MAPKK Wis1 and Sp1 MAPK as final receptor. Wis4/Win1 and Msc4 form a complex with the GapDH isoform Tdh1. H_2O_2-mediated oxidative modification of Tdh1's active site Cys152 enhances the interaction between Tdh1 and the Msc4 response regulator, which, in turn, promotes the interaction and phosphorelay signaling between the response regulator Msc4 and the HPt protein Mpr1. This interaction is required for the transmission of the H_2O_2 stress signal to Spc1.*

1.5.4 GapDH — One of many redox-regulated metabolic enzymes

GapDH is only one of many metabolic enzymes whose activity is regulated by oxidative thiol modifications [127]. Creatine kinase, for instance, which plays a central role in controlling cellular energy homeostasis in energy-demanding tissues such as muscle and brain, contains one highly oxidation-sensitive active site cysteine Cys283 [92]. ROS and RNS-mediated thiol modification of Cys283 leads to the reversible inactivation of creatine kinase, which appears to affect muscle performance and Ca^{2+} homeostasis [128]. This inactivation seems to be of particular importance during oxidative stress conditions encountered in ischemia, cardiomyopathy and several neurodegenerative disorders (reviewed in [129]). Similarly, both glycogen synthase and protein phosphatase 1 are readily inactivated upon treatment with oxidized glutathione [130,131]. This inactivation might explain the high glycogenolytic activity that has been observed in liver tissue treated with thiol oxidants [129]. Carbonic anhydrase 3, which maintains pH homeostasis in mammals, has been found to be sensitive to oxidative modifications of Cys186 and Cys181 [93]. Although earlier results suggested already a role of carbonic anhydrase 3 in the cellular response to oxidative stress [132], it remains to be determined whether these observed oxidative thiol modifications are involved. Interestingly, for many of these enzymes, oxidative modifications such as S-glutathionylation have been considered to be a protective measure rather than a redox-regulatory feature, with the purpose to protect active site cysteines against irreversible overoxidation and inactivation [39,73]. However, because loss of enzyme activity is inevitably connected with these oxidative thiol modifications and rapid reactivation is usually achieved upon return to reducing non-stress conditions, this argument might be purely semantic.

1.6 PTP1B — Redox regulation of eukaryotic signal transduction cascades

Reversible protein tyrosine phosphorylation and dephosphorylation are important posttranslational protein modifications involved in a variety of cellular signal transduction cascades that control metabolism, motility, cell growth, proliferation, differentiation, and survival [133,134]. While phosphorylation reactions are carried out by specific tyrosine kinases, dephosphorylation reactions are catalyzed by members of a large superfamily of protein tyrosine phosphatases (PTPs). In the human genome, 107 PTP encoding genes have been identified [133]. Even though members of the PTP superfamily show only low sequence similarities and possess different topologies [135,136], they are all characterized by an active site cysteine located within a highly conserved 11-residue motif [I/V]HCXXGXXR[S/T]. This active site cysteine exhibits a very low pKa value (~5.6 in PTP1B, ~4.7 in YOP phosphatase) [31,34]. It is present as a thiolate anion at physiological pH, which renders the catalytic site cysteine highly reactive and promotes its nucleophilic attack at a phosphotyrosine (pTyr) substrate [134,137]. This reaction results in the formation of a covalent phospho-cysteine intermediate, which is subsequently hydrolyzed by an activated water molecule [138]. The very same properties that make PTP's active site cysteine such an excellent nucleophile, also makes it highly susceptible to oxidants such as H_2O_2 and superoxide, and make the proteins highly sensitive to oxidative stress-mediated inactivation [138]. By regulating tyrosine phosphorylation-dependent signaling events using redox-mediated posttranslational modification of PTP activity, ROS can actively function as second-messenger [33,138].

1.6.1 PTP1B — Using cyclic sulfenamide formation as redox switch

PTP1B, a prototypic member of the PTP family, is widely expressed in multiple cell types and tissues including brain, liver, and skeletal muscle [139]. It is considered to be the physiological regulator of glucose homeostasis via its control of leptin sensitivity and its function as a major negative regulator of the insulin pathway [139,140]. Furthermore, PTP1B appears to have oncogenic properties in various cancer types such as in colon cancer, where it activates Src by catalyzing the dephosphorylation of the negative regulator residue Tyr530 [141].

In active PTP1B, the catalytic cysteine Cys215 is reduced and located at the base of a cleft that lies within the pTyr substrate binding pocket [134,135] (Fig. 1.5). Oxidation of Cys215 in the presence of mild oxidative conditions, such as 100 μM H_2O_2, causes the reversible formation of a cyclic sulfenamide species (Fig. 1.6). This oxidation is accompanied by major changes in the structure of the active site and leads to the inhibition of enzyme activity [38,142]. The generation of a cyclic sulfenamide bond in PTP1B probably occurs through the initial oxidation of Cys215 to a sulfenic acid intermediate. This step is followed by a nucleophilic attack of the backbone nitrogen atom of Ser216 on the sulfenated Cys215 forming cyclic sulfenamide and concomitantly releasing water [38,142]. Amino acids in the active site of PTP1B, such as the highly conserved His214, facilitate the cyclization by providing steric and/or electrostatic effects [38] (Fig. 1.5).

Upon formation of the active site sulfenamide, PTP1B undergoes dramatic tertiary structural rearrangements at its catalytic site (Fig. 1.5). These involve conformational rearrangements of both PTP and pTyr loops, which are central to catalysis and substrate recognition. In addition, converting the PTP loop into an open conformation exposes the Sγ-atom of Cys215, which might enhance the accessibility of the catalytic cysteine for reducing agents once non-stress conditions are restored [38]. Reducing agents such as GSH or DTT have been shown to reduce the active site sulfenamide, reverse the conformational changes, and fully restore the phosphatase activity [38,142]. In contrast, however, in the presence of excess H_2O_2, overoxidation of the active site cysteine to sulfinic and sulfonic acid occurs. This process is irreversible and causes permanent inactivation of the phosphatase. Because cyclic sulfenamide formation has been found to be stable for more than 5 hours in 50 μM H_2O_2 [38], it is thought that this oxidative thiol modification functions as a protective mechanism to prevent PTP1B against irreversible overoxidation [38,142]. A similar protection against overoxidation via sulfenamide bond formation was also shown for OhrR in *Bacillus subtilis* [143]. In addition to sulfenamide bond formation, glutathionylation of the catalytic cysteine in PTP1B might also protect the enzyme against overoxidation [144]. So far, what different roles glutathionylation and sulfenamide bond formation might play in the protection of the catalytic cysteine remains unclear. Salmeen *et al.* [145] suggest that the formation of a sulfenamide bond may be the first step in protection, followed by glutathionylation, which can occur as the enzyme is being reduced (Fig. 1.6).

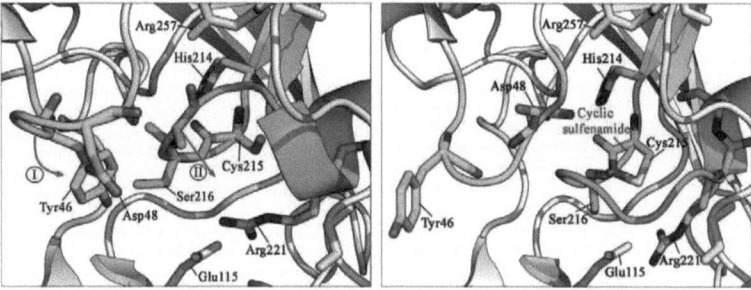

Figure 1.5. Redox-mediated conformational changes in PTP1B's catalytic site. *Shown are the reduced form of PTP1B (left panel) and the sulfenamide species of PTP1B (right panel). Oxidation and sulfenamide bond formation of PTP1B's* active site *cysteine Cys215 causes significant conformational changes in the catalytic site of PTP1B. Large rearrangements of both pTyr loop (I, left loop) and PTP loop (II, right loop) inhibit substrate binding and apparently protect the* active site *Cys215 against irreversible overoxidation. Figures were made with PyMOL using the coordinates*

of reduced human PTP1B protein (2BGE) and the sulfenamide species (1OEM) deposited in the Protein Data Bank. For molecular mechanism of sulfenamide bond formation, refer to Fig. 1.6.

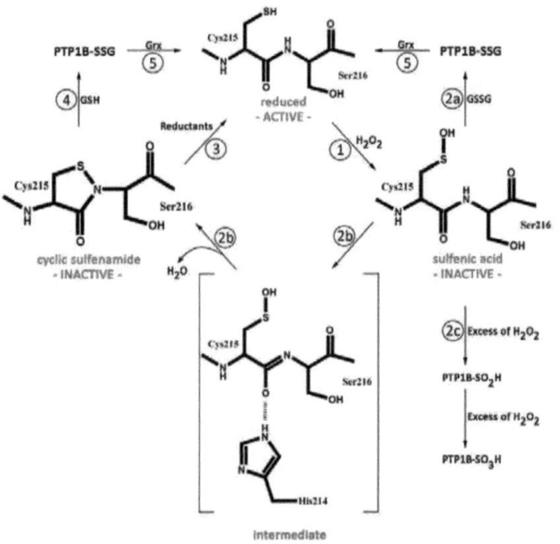

Figure 1.6. PTP1B – Redox-regulation by cyclic sulfenamide formation. Under non-stress conditions, the active site cysteine (Cys215) of tyrosine phosphatase PTP1B is reduced and PTP1B is active. *(1)* Upon exposure to oxidative stress conditions (e.g., H_2O_2), Cys215 becomes oxidized to sulfenic acid, which inactivates the enzyme. *(2a)* Reaction of the active site sulfenic acid with oxidized glutathione leads to S-glutathionylation (PTP1B-SSG). *(2b)* Alternatively, a nucleophilic attack of the backbone nitrogen of Ser216 on the Sγ-atom of Cys215 occurs, which results in the formation of a cyclic sulfenamide and the release of water. Sulfenamide formation is promoted by the environment of the catalytic site, with His214 playing a prominent role in polarizing the amide bond of Ser216. *(2c)* In the presence of excess H_2O_2, PTP1B's sulfenic acid is overoxidized to either sulfinic or sulfonic acid. *(3)* Active PTP1B can be either directly regenerated by reducing agents such as DTT in vitro or *(4)* indirectly via the reaction of cyclic sulfenamide with GSH and the formation of PTP1B-SSG. *(5)* Reduced active PTP1B is then formed by the subsequent reaction of PTP1B-SSG with the glutaredoxin (Grx) system.

1.6.2 Overoxidation of PTP1B — More than a dead-end product?

A recently applied MS-based approach revealed the extent of oxidative thiol modification of PTP1B both *in vitro* and *in vivo* [146]. This approach verified that only the active site cysteine of PTP1B is the target of thiol oxidative modifications. Surprisingly, however, quantitative analysis of the *in vivo* oxidation state of PTP1B in cancer cell lines that continuously produce ROS revealed that about 40% of PTP1B molecules were irreversibly overoxidized. Between 25-50% of all PTP1B molecules were found to be reversibly oxidized to either sulfenic acid or sulfenamide whereas only 10-15% of cellular PTP1B was found in its reduced and active form [146]. These results indicated that the level of ROS in these cell lines is sufficient to cause terminal oxidation of a significant amount of PTP1B protein, despite the presence of high levels of glutathione [147]. Interestingly, overoxidation of the active site cysteine in PTP1B appears to fully reverse the conformational changes that accompany ROS-mediated sulfenamide formation [38,142]. It is therefore not surprising that overoxidized PTP1B does not trigger the cellular machinery that is usually responsible for the rapid degradation of oxidatively damaged proteins [146]. It will now be interesting to determine whether the overoxidized form of PTP1B is indeed an end product that is destined for degradation or whether it fulfills a yet to be determined alternative function in the cell. Latter would be reminiscent of 2-Cys peroxiredoxins, which are also highly susceptible to overoxidation under high H_2O_2 concentration [148]. Sulfinic acid formation in 2-Cys peroxiredoxins leads to the inactivation of their peroxidase activity, but at the same time, appears to convert 2-Cys peroxiredoxins into proteins with molecular chaperone activity [149]. This observation then raises the question as to whether PTP1B-specific sulfinic acid reductases exist, which, similar to sulfiredoxin in the case of overoxidized 2-Cys peroxiredoxins [45], return PTP1B into its original redox and functional state.

1.6.3 Other redox-regulated signal transduction cascades in eukaryotes

Many physiological ligands such as hormones, growth factors, and cytokines trigger intrinsic ROS production. ROS-mediated regulation of protein tyrosine phosphatases (PTPs) appears therefore to represent a general mechanism by which ligands control PTP activity and signal transduction cascades. ROS-induced active site thiol modifications were demonstrated for a variety of PTPs in different cell types and included formation of disulfide bonds (e.g., LMW-PTPs, Cdc25, PTEN), cyclic sulfenamide (e.g., PTP1B and RPTPα) [38,150], and a potentially stabilized sulfenic acid (e.g., VHR and LAR) [7,151,152]. The active site cysteine of the tumor suppressor PTEN (phosphatase with sequence homology to tensin), for instance, which regulates the overall activity of the PI3 kinase signaling pathway [153], is reversibly oxidized by either RNS or H_2O_2. Oxidative inactivation of PTEN leads to increased PIP_3 levels, increased Akt phosphorylation and ultimately in cell growth [153]. Reversible inactivation of MAP kinase phosphatase 3 (MKP3) upon H_2O_2 treatment, on the other hand, promotes TNFα-induced cell death [152]. Interestingly, MKP3 does not form a disulfide with a single neighboring cysteine but can use multiple cysteines distributed in both the N-terminal and C-terminal domain to trap the sulfenic acid of the active site cysteine (Cys293) as a disulfide [7]. The availability of multiple redox active cysteines, which has also been reported for other redox-sensitive proteins [154], might represent a general mechanism to prevent irreversible overoxidation of the active site sulfenic acids.

It is interesting to note at this point that ROS generated by distinct physiological stimuli often affect only a particular subset of redox-sensitive PTPs. This specificity is well-illustrated by the PDGF-induced specific oxidation of SHP-2 [151] or the insulin-induced specific oxidation of PTP1B and TC-PTP [155]. Furthermore, in-gel phosphatase assays revealed that ROS generated by cancer cells selectively affect a subset of cellular PTPs [146] and do not, as previously assumed, oxidize all redox-sensitive PTPs that are expressed. As of now, the reason behind this specificity is unknown. One explanation might be a more localized generation of ROS upon endogenous stimuli, which promotes oxidation of only those PTPs that are close to the source of ROS production. Alternatively, however, metabolites and/or proteins specific for individual signaling pathways might modulate the redox sensitivity of a subset of phosphatases, thereby making them more susceptible to ROS-mediated modification than others. Similar observations have been made with protein kinase C, whose redox regulation appears to be modulated by different retinoids [156].

1.7 Concluding remarks

Over the past ten years, a vastly increasing number of proteins have been identified that use thiol-based redox switches to quickly regulate their function, activity, or structure in response to changes in the redox homeostasis of the cell [11,56,108]. As many of these redox-regulated proteins are central players of metabolic and signal transduction pathways, this strategy ensures that most cellular processes are under tight redox control. Here, the current knowledge about three eukaryotic proteins with thiol-based redox switches, Yap1, GapDH and PTP1B is introduced and a short overview of other redox-regulated proteins similarly involved in gene expression, metabolic pathways or signal transduction is provided. We used these model proteins to illustrate the variety of mechanisms by which cysteine thiols sense and respond to reactive oxygen and nitrogen species. However, as the field expands, so does the knowledge that no singly unifying concept can be used to characterize redox-regulated proteins. While some redox-regulated proteins harbor only one oxidative stress-sensitive cysteine (e.g., GapDH, PTP1B) [38,51,144], others make use of cysteine clusters (e.g., Yap1) [68,70] or even redox-sensitive zinc centers (e.g., Hsp33, RsrA) [29,157]. The nature of the oxidative thiol modifications depends largely on the type of oxidant as well as on the environment of the cysteine within the protein. Moreover, because most oxidative modifications affect cysteines in structurally or functionally important regions, variations in the type of oxidative modifications often create distinct conformational changes, which can result in different functional states. Finally, while only a few proteins harbor generally oxidative stress-sensitive cysteines (e.g., GapDH), most proteins have been found to be highly specific for distinct oxidative or nitrosative stressors [158]; the factors that are important for this specificity have yet to be determined.

Oxidative and nitrosative stress is involved in many disease conditions, and numerous redox-regulated proteins have been identified that use their redox-sensitivity to protect cells and organisms against these stress conditions. Despite the tremendous increase in the number of redox-regulated proteins identified, it is still unclear what makes cysteines particularly sensitive to specific oxidants in some proteins and not in others. Our challenge for the future will be to obtain a comprehensive view about what makes individual proteins redox-sensitive. Can we predict, based on sequence and structure similarities, the redox sensitivity of individual cysteines, and more importantly, a protein's potential of being redox-regulated? This capability would greatly facilitate efforts to identify proteins that are potentially involved in the oxidative stress protection of the cell. A number of global proteomic techniques have been developed that allow the identification of redox-regulated proteins in cells and organisms [42,53,54]. With the use of highly quantitative techniques such as the recently developed OxICAT method [49], the thiol oxidation status of hundreds of different proteins can be quantified in a single experiment. These techniques in combination with extensive biochemical and structural studies will have the potential to reveal novel redox-regulated proteins. These studies will deepen our understanding of redox regulation and further confirm the importance of oxidative thiol modifications as one of the key posttranslational modifications in pro- and eukaryotic organisms.

2 Nitrosative Stress Treatment of *E. coli* Targets Distinct Set of Thiol-containing Proteins

Abstract

Reactive nitrogen species (RNS) function as powerful antimicrobials in host defense but so far little is known about their bacterial targets. In this study we set out to identify *E. coli* proteins with RNS-sensitive cysteines. We found that only a very select set of proteins contain cysteines that undergo reversible thiol modifications upon nitric oxide (NO) treatment *in vivo*. Of the ten proteins that we identified, six proteins (AtpA, AceF, FabB, GapDH, IlvC, TufA) have been shown to harbor functionally important thiol groups and are encoded by genes that are considered essential under our growth conditions. Media supplementation studies suggested that inactivation of AceF and IlvC are, in part, responsible for the observed NO-induced growth inhibition, indicating that RNS-mediated modifications play important physiological roles. Interestingly, the majority of RNS-sensitive *E. coli* proteins differ from *E. coli* proteins that harbor H_2O_2-sensitive thiol groups, implying that reactive oxygen and nitrogen species affect distinct physiological processes in bacteria. We confirmed this specificity by analyzing the activity of one of our target proteins, the small subunit of glutamate synthase. *In vivo* and *in vitro* activity studies confirmed that glutamate synthase rapidly inactivates upon nitric oxide treatment but is resistant towards other oxidative stressors.

2.1 Introduction

Nitric oxide is a lipophilic free radical gas (•NO, nitric oxide radical; denoted throughout this thesis solely as NO) with complex and diverse biological functions [9]. In bacteria, low concentrations of NO are produced during anaerobic growth when nitrate and nitrite are used as terminal electron acceptors [159]. High levels of NO, on the other hand, are toxic. This antimicrobial effect is utilized by the mammalian host defense when the inducible NO synthase of macrophages generates toxic amounts of NO in an attempt to kill off invading microorganisms [160,161]. To counteract this hazard, bacteria have developed a number of defense systems that allow them to quickly respond to nitrosative stress and to protect them against nitrosative damage [160,162].

The direct cellular effects of NO are not yet fully understood. This is in part because other, partly even more reactive NO-derived reactive nitrogen species (RNS) including peroxynitrite (ONOO$^-$), nitroxyl anion (NO$^-$) and nitrogen dioxide (NO$_2$) are breakdown products of NO and contribute to the spectrum of NO-mediated damage to cells [163]. Nevertheless, it has been clearly demonstrated that NO and NO-derived RNS modulate the activity of various enzymes [160] and damage critical cell processes including protein synthesis and DNA replication [164,165].

RNS-mediated protein modifications involve the nitration of tyrosines [166] as well as the reversible binding of NO to metal centers including the heme-iron of guanylate cyclase, the [2Fe-2S]$^{2+}$ of SoxR as well as the [4Fe-4S] cluster of FNR [167-169]. Moreover, in the presence of oxygen, endogenously produced NO or NO released from compounds such as DEA/NO appear to rapidly cause S-nitrosylation of protein thiols (SNO) and glutathione (GSNO) presumably via the formation of RNS such as N$_2$O$_3$, NO$_2$, and thiyl radicals [10]. GSNO, which has been described as the most abundant source of nitric oxide in the cell [170] causes the direct formation of S-nitrosothiols in proteins. S-nitrosothiols in proteins are then often hydrolyzed to highly reactive sulfenic acids, which, in turn, are either stabilized or react with reduced glutathione to become glutathionylated or form disulfide bonds with nearby cysteines [171]. One demonstration for the multifaceted effects of nitric oxide on protein thiols came from studies with glyceraldehyde-3-phosphate dehydrogenase (GapDH). The active site cysteine of GapDH can become S-nitrosylated, oxidized to sulfenic acid or glutathionylated upon treatment with GSNO, peroxynitrite or DEA/NO, respectively [94,172]. Other targets of RNS-mediated thiol modifications are cysteine residues of antioxidant enzymes (e.g., catalase and glutathione peroxidase), kinases (e.g., Akt/PKB), proteases (e.g., caspase) and transcription factors (e.g., HIF-1) (reviewed in [170]). Depending on the individual proteins, these thiol modifications either lead to the activation or inactivation of the respective protein function.

The few studies that have so far been conducted to identify proteins with RNS-sensitive cysteines used tools to specifically enrich for and identify proteins that harbor S-nitrosylated cysteines [173]. These studies potentially miss, however, all those proteins, in which S-nitrosothiols are hydrolyzed and further modified. We therefore decided to use our recently established differential thiol trapping technique, which allows us to identify proteins, whose cysteines undergo a wide range of different reversible oxidative thiol modifications in response to nitric oxide treatment. By using this technique, we identified 10 different *E. coli* proteins that harbor RNS-sensitive thiol groups. Interestingly, the identified proteins are largely distinct from bacterial proteins that harbor H_2O_2 sensitive cysteines. They included a large proportion of genes that are essential for *E. coli* growth under our cultivation conditions. Media supplementation studies suggested that RNS-mediated modifications of AceF as well of enzymes in the branched chain amino acid biosynthesis pathway are responsible for the observed growth defect of NO-treated *E. coli* cells. To characterize the specificity of oxidative modifications in greater detail, we focused our subsequent studies on one of our identified proteins, the small subunit of glutamate synthase. *In vivo* activity assays revealed that glutamate synthase is reversibly inhibited by nitric oxide treatment but not by other reactive oxygen species such as H_2O_2 or paraquat. Similar results were obtained with the purified enzyme indicating that glutamate synthase and probably many of our other identified proteins contain cysteines, which are specifically sensitive to treatment with RNS.

2.2 Material and Methods

2.2.1 Bacterial strains and culture conditions

The *E. coli* strain *K-12 MG1655* (F- λ- ilvG- rfb-50 rph-1) was used for differential thiol trapping and enzyme assays. The deletion strains *MG1655 ΔgltD* (JW3180) was obtained from the University of Wisconsin *E. coli* Genome Project [174]. Bacterial cultures were grown at 37°C either in Luria-Bertani (LB) medium or in MOPS minimal medium [175] containing 10 µM thiamine. For media supplementation studies, *MG1655* was grown in

MOPS minimal medium using 0.2% glucose or 0.6% acetate as carbon source. To supplement the media with branched chain amino acids, 0.35 mM Ile, 0.35 mM Leu and 0.3 mM Val was added. Media for plasmid selection and maintenance were supplemented with ampicillin (200 µg/ml) or kanamycin (25 µg/ml). Nitric oxide solutions were freshly prepared by either dissolving diethylamine NONOate (DEA/NO) in 10 mM KOH or S-nitroso-L-glutathione (GSNO) in H_2O (both from Cayman Chemicals, Ann Arbor). The stability of DEA/NO in growth media was monitored by its characteristic UV absorbance at 250 nm [176]. No influence of the culture conditions on the DEA/NO stability was observed. H_2O_2 solutions and paraquat were obtained from Fisher Scientific and MP Biochemicals, respectively.

2.2.2 Plasmid construction

Plasmids were constructed using standard recombinant DNA techniques. PCR reactions were performed using Pfu Turbo DNA polymerase according to the manufacturer's manual and DNA was purified using the PCR Purification Kit (Quiagen). Plasmids were isolated using the SV Miniprep plasmid kit (Promega). All DNA modifying enzymes were from Promega. The 6-kb coding region of *E. coli gltB*, encoding the α-subunit of GOGAT and *gltD*, which encodes the β-subunit of GOGAT, was amplified by PCR from *E. coli K-12 MG1655* genomic DNA using the primers 5'-CGTTCTAGAATGACACGCAAACCC-3' and 5'-ATAGGATCCACTCATTAAACTTC CAGCC-3', respectively, which incorporated XbaI and BamHI restriction sites. The amplified genes were cloned into the expression vector pET11a (Novagen), thereby creating plasmid pET11gltBD. The correct sequence of the plasmid was confirmed by sequencing.

2.2.3 Differential thiol trapping

2.2.3.1 Sample preparation and differential thiol trapping

MG1655 E. coli cells were grown at 37°C in MOPS minimal media until an OD_{600} of 0.4 was reached. The translation inhibitor chloramphenicol (200 µg/ml) was added to the bacterial cell culture 10 min prior to the stress treatment. The cell culture was split and treated with different concentrations of nitric oxide ranging from 0 to 0.5 mM DEA/NO. After 10 min of incubation at 37°C, 1.8 ml of the cell culture was harvested directly onto 200 µl of TE-buffer supplemented with 1 M iodoacetamide (IAM). Then, the thiol trapping protocol established by Leichert and Jakob [42] was followed with slight modifications, which concerned mainly the avoidance of trichloroacetic acid (TCA) during the trapping procedure. This was necessary to prevent uncontrolled nitric oxide formation during sample preparation, which can be generated when nitrite, one of the possible end products of nitric oxide treatment is acidified [177]. Cells were pelleted by centrifugation (5,000 x g, 4°C, 1 min) and resuspended in 100 µl of denaturing buffer (6 M Urea, 200 mM Tris-HCl pH 8.5, 10 mM EDTA, 0.5% (w/v) SDS) supplemented with 100 mM IAM. This first alkylation procedure irreversibly modified all free thiol groups that were made accessible by urea and SDS-denaturation of the proteins. After 10 min of incubation at 25°C, the reaction was stopped by adding 500 µl of ice-cold 100% acetone. After incubation at -20°C for 2 h, the precipitated and alkylated proteins were pelleted (13,000 x g, 4°C, 30 min) and the protein pellet was washed with 80% acetone.

The protein pellet was then dissolved in 20 µl of 10 mM DTT in denaturing buffer to reduce all reversible thiol modifications such as disulfide bonds and sulfenic acids. After 1 h at 25°C, 20 µl of a solution of 100 mM radioactively labeled [^{14}C]-IAM (PerkinElmer) in denaturing buffer was added to titrate out the DTT and to irreversibly alkylate all newly reduced cysteines. The reaction mixture was incubated for 10 min at 25°C. The reaction was stopped by adding 40 µl of 20% (w/v) TCA. After precipitation on ice and subsequent centrifugation, the pellet was washed first with TCA and then three times with 500 µl of ice-cold ethanol. For protein identification purposes, thiol-trapping experiments using nonradioactive IAM in both alkylation steps were performed in parallel 2D gel electrophoresis. The pellet of the thiol-trapped proteins was dissolved in 500 µl of rehydration buffer (7 M urea, 2 M thiourea, 1% (w/v) Serdolit MB-1, 1% (w/v) dithiothreitol, 4% (w/v) Chaps, and 0.5% (v/v) Pharmalyte 3–10), and the 2D gel electrophoresis was performed as previously described [178].

2.2.3.2 Staining of the gels, storage phosphor autoradiography, and image analysis

Gels were stained using colloidal Coomassie blue stain and scanned using an Expression 1680 scanner with transparency unit (Epson America) at 200 dpi resolution and 16-bit grayscale. Phosphor images were obtained by exposing LE Storage Phosphor Screens (Amersham Biosciences) to dried gels for 7 d. The phosphor image screens were read out with the Personal Molecular Imager FX (Biorad) at a resolution of 100 µm. The original image size of the phosphor image was changed to a resolution of 200 dpi with PhotoShop

7.0 (Adobe Systems). The phosphor images and images of the stained proteins were analyzed using Delta 2D Software (Decodon).

2.2.3.3 Data analysis and identification of proteins from 2D gels

For each of the described experiments, at least 2 individually trapped samples of cultures were prepared from at least 2 independent cell cultures. For each of the experiments, the phosphor image with the highest overall ^{14}C-activity was chosen for spot detection. The 100 most abundant spots were chosen from the detected set of spots and the boundaries transferred to all other phosphor images and protein gel images using the Delta 2D "transfer spots" function. The absolute intensity for each of these 100 spots on the protein gels and the phosphor image was determined to quantitatively describe the amount of protein and ^{14}C-activity for each protein spot. These absolute spot intensities were then normalized over all 100 spots. Finally, the ratio of ^{14}C-activity/protein was calculated by dividing the normalized intensity of the protein spot on the phosphor image by the corresponding normalized intensity of the Coomassie blue-stained protein spot. For a protein to be considered significantly thiol-modified, the average of the ^{14}C-activity/protein ratio for a given protein spot had to be at least 1.4-fold above the average of the ^{14}C-activity/protein ratio of this protein under control conditions. Thiol trapped samples using non-radioactive IAM in both alkylation steps were separated on 2D gels and used to excise proteins of interest. These proteins were identified by Peptide Mass Fingerprinting at the Michigan Proteome Consortium (http://www.proteome consortium.org).

2.2.4 Purification of Glutamate Synthase (GOGAT)

E. coli BL21(DE3) cells were transformed with the plasmid pET11gltBD and grown aerobically at 37°C in LB medium containing 0.2 mg/ml ampicillin and 0.05 mg/ml kanamycin until an OD_{600} of 0.5 was reached. Then, expression of the two glutamate synthase subunits GltB and GltD was induced by adding 1 mM isopropyl thio-β-D-galactoside (IPTG). Two hours after IPTG induction, cells were harvested by centrifugation (4,000 × g, 4°C, 20 min) and resuspended in 20 mM of potassium phosphate buffer, 1 mM EDTA pH 7.6 at a ratio of 1 g of cells per 2 ml of buffer. All following steps were carried out at 4°C. The purification of GOGAT was based on a protocol described by Hashim et al. [179] with slight modifications that will be reported elsewhere.

2.2.5 Enzyme assays

MG1655 was grown at 37°C in MOPS minimal medium to OD$_{600}$ of 0.5 and treated with the indicated concentrations of oxidants (DEA/NO, GSNO, H$_2$O$_2$, paraquat) for 10 min. Cells were harvested by centrifugation (6,000 x *g*, 5 min, RT), washed twice with 1% (w/v) KCl, and resuspended at 10-fold concentration in 1% KCl [180]. The concentrated cell extract was used for GOGAT and IlvC *in vivo* activity assays, which were performed as previously described [181,182]. Both activities were assayed at 30°C by measuring the initial rates of NADPH oxidation at 340 nm. For GOGAT activity assays, 50 µl of concentrated cells were added to the reaction cocktail (0.15 mM NADPH, 50 mM Tris pH 7.6, 5 mM α-ketoglutarate, 0.1% (w/v) cetyl trimethyl ammonium bromide (CTAB), 0.01% sodium deoxycholate (NaDOC) in 0.8 ml of H$_2$O). Reactions were started by addition of 5 mM L-glutamine for GOGAT activity measurements. The specific activities are expressed as nanomoles of NADPH oxidized per minute per milligram of protein. To determine IlvC activity, 200 µl of concentrated cells were added to the reaction cocktail (0.2 mM NADPH, 100 mM Tris pH 7.6, 10 mM MgCl2, 1 mM acetolactate, 0.1% (w/v) CTAB, 0.01% NaDOC in 0.7 ml of H$_2$O). Racemic 2-acetolactate was prepared by alkaline hydrolysis of 2-hydroxy-2-methyl-3-ketobutyrate (Sigma-Aldrich) [183], and concentration of synthesized acetolactate was determined according to the method of Westerfeld [184]. Reactions were started by addition of 0.2 mM NADPH. To determine the influence of oxidants on the *in vitro* activity of purified GOGAT, 0.2 µM GOGAT in 20 mM potassium phosphate buffer pH 7.6 was incubated with the indicated concentrations of DEA/NO, GSNO or H$_2$O$_2$ at 30°C. At defined time points, 50 µl of this solution was mixed with the reaction cocktail (excluding NaDOC and CTAB) to a final GOGAT concentration of 10 nM. 5 mM L-glutamine was added to start the reaction and the initial rates of NADPH oxidation were measured as described above.

2.2.6 UV absorbance and analysis of GOGAT's thiol status using mass spectrometry

To prepare oxidized GOGAT, 4 µM purified GOGAT in 20 mM PPB was incubated with 1.5 mM DEA/NO for 15 min at 30°C. Then, DEA/NO was removed using a NAP5 column (Amersham Biosciences). The UV/VIS spectra were recorded and normalized for the absorption at 280 nm using a CARY 100 Bio UV/VIS spectrophotometer (Varian Instruments, USA) interfaced to a computer. To analyze the thiol status of GOGAT, 30 µl of a 4 µM solution of purified GOGAT or GOGAT oxidized with 0.5 mM DEA/NO for 15 min at 30°C was precipitated with 10% (w/v) TCA [29]. The pellets were resuspended in 20 µl of DAB buffer (6 M urea, 200 mM Tris-HCl pH 8.5, 10 mM EDTA, 0.5% (w/v) SDS) containing 100 mM NEM for 1 h under vigorous shaking. The proteins were then precipitated again with 10% (w/v) TCA and resuspended in 20 µl DAB buffer supplemented with 10 mM DTT to reduce all existing oxidative thiol modifications. After 1 h at 25 °C, 20 µl of 50 mM IAM was added and all newly accessible thiol groups were alkylated for 1 h at 25 °C. All small molecules were removed with C4–reversed phase material deposited in gel-loader pipette tips (Millipore). Tryptic digests of GOGAT and MALDI-MS analysis of the peptides were conducted at the Michigan Proteome consortium.

2.3 Results and Discussion

2.3.1 Global analysis of the thiol-disulfide state of *E. coli* proteins upon DEA/NO treatment

Reactive nitrogen species (RNS) are potent antimicrobial agents released by host immune cells upon stimulation [160,161]. To identify potential RNS-sensitive target proteins in *E. coli*, we applied a differential thiol trapping technique that we recently developed [42]. To determine the concentrations of nitric oxide that affect the growth of *E. coli*, we cultivated wild type *E. coli MG1655* in minimal media. We then exposed exponentially growing bacteria to increasing concentrations of the NO-donor diethylamine NONOate (DEA/NO) and further monitored cell growth at 37°C (Fig. 2.1). Under these temperature and pH conditions, the half-life of DEA/NO is 2 minutes and DEA/NO liberates 1.5 moles of NO per mole of parent compound [176]. We found that DEA/NO treatment transiently affected the growth rate of *E. coli* without killing the cells (Fig. 2.1). This result served as good indication that the original redox status of proteins and cellular compartments can be restored over time and excluded the possibility that we are following the effects of oxidative stress-induced cell death. Our results were in good agreement with previous studies that analyzed the transcriptional effects of DEA/NO treatment in bacteria [185].

In contrast to the studies by Hyduke and co-workers, however, in which 8 µM DEA/NO was sufficient to induce pronounced bacteriostasis, at least 5-fold higher DEA/NO concentrations were needed in our studies to substantially affect cell growth. These differences are likely due to differences in media composition, which might influence the effective NO concentration that diffuses into bacteria and/or due to the different strain backgrounds that were used in these studies.

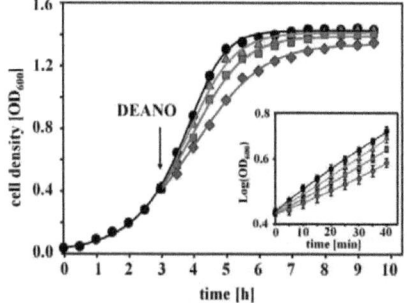

Figure 2.1. Influence of DEA/NO-treatment on *E. coli* growth. MG1655 cells were grown at 37°C in MOPS minimal medium to OD_{600} of 0.4 and treated either with 0 mM DEA/NO (●), 10 µM DEA/NO (▲), 50 µM DEA/NO (■) or 0.5 mM DEA/NO (♦) at 37°C. Inset: E. coli growth during the first 40 min after addition of increasing concentrations of DEA/NO is shown.

To analyze the effects of nitric oxide treatment on the thiol status of the *E. coli* proteome, we treated *E. coli* with 0.05 mM and 0.5 mM DEA/NO and removed aliquots before as well as 10 min after the DEA/NO treatment. We then utilized a slightly modified differential thiol trapping technique, which is based on the sequential reaction of two variants of iodoacetamide (IAM) with accessible cysteine residues in proteins [42]. In the first step of this labeling method, all reduced cysteines in proteins are irreversibly carbamidomethylated with cold IAM. In a second step, all reversible thiol modifications including S-nitrosothiols, disulfide bonds and sulfenic acids are reduced using DTT. Then all newly accessible cysteines are modified using ^{14}C-labeled IAM. At the end of this trapping protocol, the thiol groups in all proteins are irreversibly carbamidomethylated but only those proteins that originally contained thiol modifications will be radioactive allowing their identification and quantification. The proteins are then separated on 2D gels and the extent of ^{14}C-incorporation is determined by autoradiography. High ratios of ^{14}C-activity/protein occur in proteins with *in vivo* thiol modifications while low ratios of ^{14}C-activity/protein are found in proteins whose thiol groups are not significantly modified [42].

Analysis of the 100 most thiol modified proteins in *E. coli* cells treated with 0.5 mM DEA/NO for 10 min (Fig. 2.2) showed that only 10 proteins responded to these severe nitrosative stress conditions with a more than 1.4-fold increase in their thiol oxidation state. Most other thiol modified proteins are periplasmic proteins, which are fully oxidized in aerobically growing *E. coli* cells (e.g., OmpA, DsbA) and whose thiol groups are not affected by nitric oxide treatment (Fig. 2.2) [42]. The extent of thiol modifications in the 10 proteins that apparently harbor RNS-sensitive thiol groups was in each case dependent on the DEA/NO concentration used (Table 1.1).

Importantly, we noted that a large proportion (60%) of our identified proteins have been described to be conditionally essential for *E. coli* under the conditions used in this study, possibly explaining the antimicrobial effects of NO treatment [186,187]. This result was in excellent agreement with a recently conducted nitroso-proteome study in the Gram-positive *Mycobacterium tuberculosis* [173]. Upon treatment of *M. tuberculosis* with millimolar concentrations of the NO-donor acidified nitrite, 29 S-nitroso proteins were identified. 62% of these proteins were found to be encoded by genes that are predicted to be essential or required for optimal growth of *M. tuberculosis* [173].

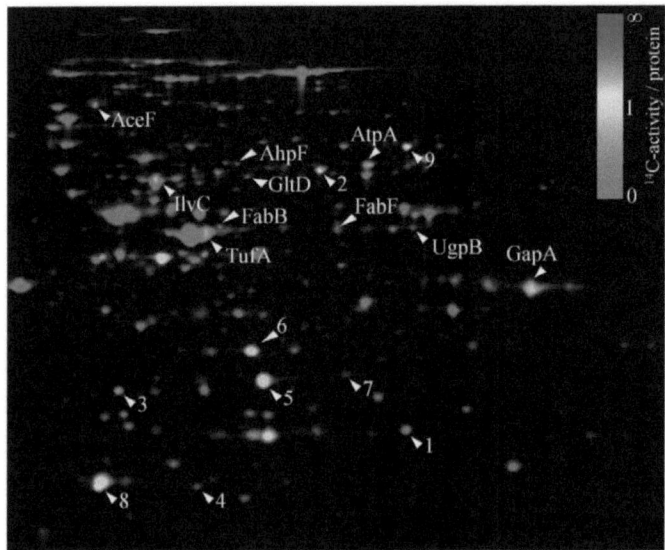

Figure 2.2. DEA/NO treatment affects the thiol-state of only a selected group of *E. coli* proteins. *False colored overlay of the Coomassie blue-stained 2D gel (shown in green) and the autoradiograph (shown in red) of a differentially thiol-trapped protein extract from exponentially growing E. coli cells 10 min after treatment with 0.5 mM DEA/NO. The color code represents the relative oxidation status of proteins, with green spots representing proteins, whose thiol groups are largely reduced in vivo while red spots represent proteins, whose thiol groups are oxidized in vivo. The ten labeled* E. coli *proteins have a more than 1.4-fold higher ratio of ^{14}C-radioactivity/protein in DEA/NO-treated cells as compared to untreated control cells and are considered to harbor NO-sensitive thiol groups. A selected group of proteins, which are already oxidized under control conditions are used as reference and indicated with a number: (1) ArtJ, (2) DppA, (3) HisJ, (4) DsbA, (5) OmpA IF 1, (6) OmpA IF 2, (7) OmpA IF 5, (8) AhpC, (9) OppA.*

Table 2.1 Nitrosative stress sensitive proteins in *E. coli*

Gene	Protein	# of Cys	Fold change in ^{14}C-activity/protein after 10 min of DEA/NO treatment[a]		Location of Cysteines
			0.05 mM	0.5 mM	
aceF	Dihydrolipoamide S-acetyltransferase	1	1.6	1.7	Prosthetic group: thiol-containing lipoamide [188]
ahpF	Alkyl hydroperoxide reductase, large subunit F	6	1.7	2.3	Two redox-active disulfides per subunit (Cys129/Cys132 and Cys345/Cys348) [189]
atpA	ATP synthase subunit alpha	4	1.1	1.6	Cys90 in conserved patch [190]
fabB	3-oxoacyl-(acyl carrier protein) synthase I	6	1.3	2.0	Cys63 in active site [191]
fabF	3-oxoacyl-(acyl carrier protein) synthase II	4	1.2	1.7	Cys164 in active site [192]
gapA	Glyceraldehyde-3-P dehydrogenase	3	1.2	1.9	Cys149 in active site [193]
gltD	Glutamate synthase, small subunit	11	1.1	1.9	Two [4Fe-4S]$^{1+,2+}$ clusters [194]
ilvC	Ketol-acid reductoisomerase	6	0.9	1.4	Cys45/Cys156 in conserved patch, Cys226 highly conserved [190]
tufA/B	Translation elongation factor Ef-Tu	3	1.0	1.5	Cys81 associated with aminoacyl-tRNA binding; Cys137 associated with GTP binding [195]
ugpB	sn-glycerol 3-phosphate transport protein	2	1.1	1.6	Cys264 in conserved patch [190]

2.1. Nitrosative stress sensitive proteins in *E. coli*. [a]*The activity/protein ratio of the given protein spot on gels from differentially trapped extracts from E. coli MG1655 (wild type) treated with different concentrations of DEA/NO for 10 min was divided by the corresponding ratio on gels from differentially trapped extracts from untreated E. coli MG1655. The data presented are the average of two different experiments, each of which included two separate samples. The experiments were performed in the presence of chloramphenicol to block de novo protein synthesis.*

2.3.2 NO-treatment causes oxidative thiol modification in a distinct set of *E. coli* proteins

Three of our identified proteins, glyceraldehyde-3-phosphate dehydrogenase (GapDH), the alpha subunit of ATP synthase (AtpA) and the elongation factor Tu (Ef-Tu) have been previously shown to harbor RNS-sensitive cysteines [173,196] (Table 2.1). While the reported thiol modifications of GapDH's active site cysteine included S-nitrosylation, sulfenic acid formation or glutathionylation depending on the type of RNS used [94,172], AtpA and Ef-Tu were both identified to be S-nitrosylated upon treatment of *M. tuberculosis* with acidified nitrite [173]. To investigate the nature of our observed thiol modifications, we applied the biotin switch method, which was designed to identify nitrosylated proteins in cell lysates [197].

We did not obtain any conclusive results with the biotin switch method possibly for the reasons described elsewhere [198,199]. Thus, we developed a variation of our differential thiol trapping technique, in which we used ascorbic acid instead of DTT to specifically reduce and then label S-nitrosylated cysteines in our DEA/NO-treated samples with ^{14}C-IAM. Ascorbic acid has been previously used to reduce nitrosothiol groups [197]. We were unable to detect any significantly S-nitrosylated proteins in our DEA/NO-treated *E. coli* cells (data not shown). This can either be due to the lack of suitable methods that are able to detect S-nitrosylated proteins or might indicate that most S-nitrosylations are unstable and rapidly convert to other oxidative thiol modifications either *in vivo* or during our sample preparation *in vitro*. Because the biotin switch method is based on the specific biotinylation and purification of only those proteins, whose cysteines are S-nitrosylated and therefore sensitive to ascorbic acid treatment, it enriches for S-nitrosylated proteins [197]. This, however, makes it impossible to quantify the proportion of proteins that harbor S-nitrosothiols. A recently developed computer program, which simulates metal-independent RNS chemistry suggests that nitrosothiols might not be the end product but rather intermediates that only transiently accumulate *in vivo* [200]. Our data agree with this simulation and suggest that the majority of our identified RNS-sensitive proteins do not contain S-nitrosothiols but other reversible oxidative thiol modifications.

All remaining proteins that came out of our screen appear to be newly identified bacterial RNS-target proteins. One of these proteins is dihydrolipoyl transacetylase AceF, the E2 component of pyruvate dehydrogenase complex (PDC) [201]. Reactive nitrogen species most likely modify the two vicinal thiol groups in the lipoyl group of AceF, which have been previously suggested to be the target of environmental toxins such as arsenic [202]. RNS treatment of *M. tuberculosis* has recently been shown to lead to the nitrosylation of dihydrolipoyl dehydrogenase (Lpd), the E3 component of PDC, whose two active site cysteines are essential for the reoxidation of AceF's lipoyl group [173]. We were unable to detect Lpd as significantly oxidized protein in DEA/NO treated *E. coli* cells.

Another RNS-sensitive protein, which is encoded by a conditionally essential *E. coli* gene, is the 3-oxoacyl-(acyl carrier protein) synthase I (FabB). FabB plays a critical role in the elongation cycle of type II fatty acid biosynthesis. Like the closely related 3-oxoacyl-(acyl carrier protein) synthase II (FabF), which we also identified as RNS-sensitive target protein (Table 2.1), FabB uses one highly conserved cysteine for its catalytic activity. Modification of this active site cysteine in both FabB and FabF would substantially interrupt fatty acid synthesis in bacteria as well as impact the thermoregulation of membrane fluidity [203].

The ketol-acid reductoisomerase IlvC was the last protein in our subset of RNS-sensitive proteins, which are encoded by conditionally essential *E. coli* genes. IlvC plays a central role in the biosynthesis of isoleucine and valine [187]. Targets of oxidative thiol modification in IlvC could either be a cysteine pair (Cys145/Cys156) located within a highly conserved region of the sequence and/or a very highly conserved single Cys226 [190]. Noteworthy, dihydroxyacid dehydratase IlvD, which uses the reaction product of IlvC as its substrate in the Ile/Val biosynthetic pathway, has only recently been suggested to contain a NO-sensitive [Fe-S] cluster [185].

Alkylhydroperoxide reductase AhpF was one of the remaining proteins that we found to be increasingly thiol modified in response to nitric oxide treatment. AhpF acts as dedicated reductase for alkylhydroperoxidase AhpC, which uses disulfide bond formation to detoxify alkyl hydroperoxides in *E. coli* [204]. While the C-terminus of AhpF contains the FAD/NAD binding domain, the N-terminal domain is homologous to protein disulfide isomerase (PDI), a protein that is involved in the oxidative protein folding in the endoplasmic reticulum [205]. In eukaryotes, PDI has recently been shown to have S-denitrosation activity, which in turn leads to disulfide bond formation in both of the active site CXXC motifs of PDI [206]. Our finding that AhpF accumulates in the oxidized state while no apparent changes were detected in the oxidation status of AhpC suggested that *E. coli* AhpF might have AhpC-independent denitrosation activity similar to that found in PDI.

Finally, we identified the glycerol 3-phosphate transport protein UgpB as well as the small subunit (GltD) of glutamate synthase to be highly RNS-sensitive *E. coli* proteins. GltD turned out to be the only [Fe-S] cluster-containing protein in our set of proteins that harbor RNS-sensitive thiol groups. GltD contains at total of 11 cysteines, of which 8 are proposed to be involved in the formation of two oxidation-stable [4Fe-4S]$^{1+,2+}$ clusters [207]. That only one of our identified RNS-sensitive target proteins contain [Fe-S] clusters was remarkable, given that [Fe-S] clusters have been found as targets of reactive nitrogen species in several other proteins [169,208]. This result might reflect, however, the preference of nitric oxide to oxidize the iron in the cluster rather than the cysteines that coordinate the clusters, an event that we would not detect using our differential thiol trapping method.

2.3.3 Identification of essential RNS-sensitive proteins involved in DEA/NO-induced growth inhibition

A significant proportion of our RNS-modified proteins are encoded by genes that have previously been shown to be essential for *E. coli* under our growth conditions used [186,187]. To investigate whether RNS-mediated thiol modification of one or more of these essential proteins is responsible for the observed growth arrest, we performed media supplementation experiments. To investigate whether the growth inhibitory effects of DEA/NO treatment are caused by the transient inactivation of AceF, one of the three enzymes constituting the pyruvate dehydrogenase complex, for instance, we analyzed the DEA/NO sensitivity of *MG1655* in MOPS minimal media supplemented with 0.6% acetate as carbon source. The PDC catalyzes the oxidative decarboxylation of pyruvate to acetyl-CoA, a key reaction that connects glycolysis with the TCA cycle. Inactivation of any of the three PDC components leads to the cessation of flux to acetyl-CoA and causes growth inhibition on glucose. Growth on acetate, which can be directly converted to acetyl-CoA, allows *E. coli* to bypass this key reaction provided that gluconeogenic enzymes such as GapDH are available to form C6-precursor molecules. Therefore, DEA/NO treatment of *MG1655* in media supplemented with acetate should no longer affect *E. coli* growth if RNS-induced inactivation of AceF was indeed the major cause of the observed DEA/NO-induced growth inhibition. Likewise, we supplemented media with branched chain amino acids to investigate the role of IlvC's thiol modification in DEA/NO-induced growth inhibition. Supplementation with branched chain amino acids has recently been shown to alleviate the DEA/NO-induced growth arrest in BW25113 cells [185].

Like before, we exposed the cells to the various concentrations of DEA/NO and monitored their growth. As shown in Fig. 2.3B, *E. coli* cells grown in media supplemented with 0.6% acetate as carbon source were no longer affected by 10 μM DEA/NO and were reproducibly less affected by higher concentrations of DEA/NO than *E. coli* cells grown in glucose-containing media (Fig. 2.3A). This result suggests that the DEA/NO-induced changes in *E. coli* growth observed at low concentrations of DEA/NO are caused by the inactivation of PDC and largely excluded a major involvement of GapDH's oxidative modification. This result was in excellent agreement with our *in vivo* thiol trapping data, which revealed that AceF shows one of the highest degrees of oxidative thiol modification of any protein at these low DEA/NO concentrations (Table 2.1). Supplementation with branched chain amino acids Ile/Leu/Val was found to reduce the growth inhibitory effect of DEA/NO treatment, although it was unable to fully prevent the growth inhibition especially at high DEA/NO concentrations (Fig. 2.3C). *E. coli* cells cultivated in media supplemented with both acetate and branched chain amino acids, on the other hand, were completely resistant to DEA/NO concentrations of up to 50 μM (Fig. 2.3D). We concluded from these results that oxidative modification of PDC as well as of enzymes of the branched chain amino acid biosynthesis contribute to the growth inhibitory effect that low micromolar concentrations of DEA/NO exert on *E. coli* cells grown in MOPS minimal media. Interestingly, media supplementation with acetate and branched chain amino acids was not sufficient to prevent growth inhibition at 0.5 mM DEA/NO. This result suggested that under these extreme nitrosative stress conditions, modification of other protein(s) such as Ef-Tu or ATP synthase might become responsible for the inhibition of *E. coli* growth.

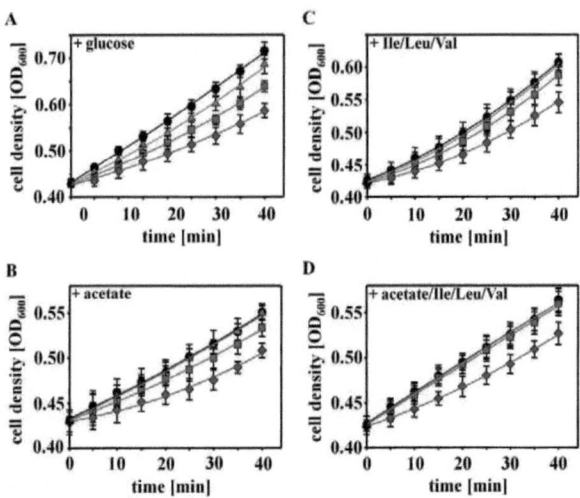

Figure 2.3. Identification of essential proteins involved in DEA/NO-induced growth inhibition. E. coli MG1655 *wild type cells were grown in MOPS minimal medium supplemented with* **(A)** *0.2% glucose,* **(B)** *0.6% acetate,* **(C)** *0.2% glucose and Ile/Leu/Val (0.35 mM/0.35 mM/0.3 mM) or* **(D)** *0.6% acetate and Ile/Leu/Val until* OD_{600} *of 0.4 was reached. The cultures were split and treated either with 0 mM DEA/NO (●), 10 μM DEA/NO (▲), 50 μM DEA/NO (■) or 0.5 mM DEA/NO (♦) at 37°C. The mean and standard deviation of three independent experiments is shown.*

2.3.4 Activity of IlvC is affected by DEA/NO treatment *in vivo*

E. coli cells cultivated in the absence of branched chain amino acids were significantly more sensitive to micromolar concentrations of DEA/NO than E. coli cells grown in their presence (compare Fig. 2.3, A and C). This result has been observed before and was attributed to the inactivation of the [Fe-S] containing dihydroxyacid dehydratase IlvD, an enzyme essential in the Ile/Val biosynthesis pathway [185]. This conclusion was based on IlvD activity assays, which revealed that treatment of E. coli cell lysates with 8 μM DEA/NO causes a 40% decrease in IlvD activity. Furthermore, *in vivo* studies showed that IlvD overexpression prevents the growth inhibitory effect that 8 μM DEA/NO exert on BW25113 cells [185]. Our identification that IlvC, the second essential enzyme of this pathway, is substantially modified by DEA/NO treatment *in vivo*, suggested that inactivation of IlvC might also contribute to the observed growth inhibitory effect of DEA/NO. To test the effects of DEA/NO on IlvC activity *in vivo*, we treated *MG1655* with the indicated concentrations of DEA/NO. We then washed and lysed the cells and determined IlvC's enzyme activity. As shown in Fig. 2.4, IlvC activity gradually decreases with increasing concentrations of DEA/NO. This result suggests that reactive nitrogen species target two enzymes of the same amino acid pathway. RNS-mediated inactivation of these enzymes eventually depletes the pool of Ile/Val, which apparently contributes to the growth inhibition observed during DEA/NO treatment.

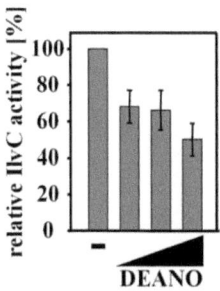

Figure 2.4. Ketol-acid reductoisomerase IlvC is sensitive to nitric oxide treatment *in vivo*. E. coli *MG1655 wild type cells were grown in MOPS minimal medium to* OD_{600} *of 0.4 and treated with increasing concentrations of DEA/NO (0.01 mM, 0.1 mM, 0.5 mM). After 10 min of incubation, aliquots were removed and tested for IlvC activity. IlvC activity in the absence of DEA/NO treatment was set to 100%. The mean and standard deviation of three independent experiments is shown.*

Interestingly, changes in IlvC's enzymatic activity were already observed at DEA/NO concentrations that did not apparently alter the thiol-disulfide status of IlvC as judged by our differential thiol trapping technique (Table 2.1). This result might indicate that functionally important residues other than cysteines are targeted by RNS as well. Alternatively, this result might illustrate the limits of the differential thiol trapping technique in proteins with numerous cysteines [42]. The differential thiol trapping technique is based on a ratio approach where the largest change in oxidation ratio is observed for proteins in which all cysteine residues are completely reduced under non-stress conditions and fully oxidized under stress conditions. Proteins that harbor cysteines that are in equilibrium with their redox environment, on the other hand, might not reveal dramatic changes in the overall oxidation ratio upon oxidative stress treatment, although the oxidation status of one individual, potentially functionally important, cysteine might be dramatically altered.

2.3.5 Glutamate Synthase — A protein specifically sensitive to nitrosative stress

Comparison of the proteins that harbor RNS-sensitive cysteines with the proteins that we have previously found to harbor H_2O_2-sensitive cysteines [42] revealed that the two sets of proteins are largely non-overlapping. H_2O_2 sensitive proteins that we previously identified included the thiol peroxidase Tpx, which is involved in detoxifying H_2O_2 *in vivo*, methionine synthase MetE, which has been shown to become rapidly glutathionylated upon H_2O_2

stress both *in vivo* and *in vitro* [209], the oxidation sensitive GTP cyclohydrolase FolE as well as the metabolic enzymes GlyA and SerA [42]. To investigate whether the identified proteins in this study have indeed thiol groups that are specifically sensitive towards thiol modifications induced by RNS, we analyzed the functional consequences of nitrosative and oxidative stress treatment on one of our identified target proteins. We decided to focus on GltD, the small (β) subunit of glutamate synthase GOGAT (i.e., NADPH-dependent glutamine:2-oxoglutarate amidotransferase), which we found to be increasingly thiol modified in response to increasing concentrations of DEA/NO (Table 2.1). GltD was the only protein from our list of RNS-sensitive target proteins that contains both, cysteines involved in [4Fe-4S] clusters and non-metal coordinating cysteines. Besides the eight cysteines that potentially coordinate the two [4Fe-4S] clusters in GltD, the protein has three additional cysteines, which we found to be conserved within a sub-group of about 60 GltD homologues [190,207].

GOGAT, which belongs to the family of eubacterial NADPH-dependent glutamate synthases, is part of the cyclic two-step glutamine synthetase (GS)-glutamate synthase pathway that synthesizes glutamate under conditions of ammonia-limited growth [210]. It catalyzes the transfer of the amido group from glutamine to α-ketoglutarate in a process that yields two molecules of glutamate [211]. The GOGAT holoenzyme is composed of four dimers, with each dimer consisting of a large (α, GltB) and small (β, GltD) subunit [212]. While the α-subunit contains the FMN cofactor and one $[3Fe-4S]^{0,+1}$ cluster, the β-subunit (GltD) contains the single FAD cofactor and two $[4Fe-4S]^{+1,+2}$ clusters. Recently it has been shown that the two [4Fe-4S] clusters are not present in isolated GltD subunits, suggesting that they are located at or near the interface between the α and β subunits [213].

To investigate whether NO-induced thiol modifications have any effect on the *in vivo* activity of GOGAT and to test whether this is specific for RNS treatment or represents a general redox-sensitivity, we determined the activity of GOGAT in *E. coli* cells treated with various RNS and ROS. Like before, we cultivated wild type *E. coli* cells to midlogarithmic growth phase and exposed them to various concentrations of the NO-donors DEA/NO and GSNO, as well as to H_2O_2 and the superoxide-generating reagent paraquat. 10 min after the treatment, we harvested the cells and measured the *in vivo* GOGAT activity. As shown in Fig. 2.5A, we found that the *in vivo* activity of GOGAT dramatically decreased in response to increasing concentrations of DEA/NO or GSNO but was essentially unchanged by H_2O_2 or paraquat treatment (Fig. 2.5A). Within 10 minutes of treatment with 0.5 mM DEA/NO, less than 20% of the original GOGAT activity was observed. Very similar results were observed with GSNO, although significantly higher concentrations were necessary to achieve similar effects. This is probably due to a slower uptake of GSNO into *E. coli* cells. While the lipophilic NO released from DEA/NO rapidly passes through *E. coli* membranes [214], GSNO relies on dipeptide permeases [215], whose expression might become limiting in minimal media [216].

We found that the sensitivity of glutamate synthase is specific for nitric oxide and RNS because neither high concentration of H_2O_2 nor of the superoxide-producing paraquat substantially affected GOGAT activity *in vivo*. These results agreed with our previous *in vivo* thiol trapping studies that did not reveal GltD as a particularly H_2O_2 sensitive *E. coli* protein [42]. Our results suggested therefore that GOGAT is a protein, whose thiol groups and functional activity are specifically sensitive to nitrosative stress agents. To ascertain that the components of our activity assay are not affected by the NO-treatment, we conducted a GOGAT activity assay using DEA/NO-treated cell lysates, which showed the expected reduction in GOGAT activity. We then added wild type lysate directly to this assay mixture and found the expected high GOGAT activity of untreated cell lysates (data not shown), indicating that DEA/NO does not affect any components of the GOGAT activity assay. These experiments showed that it is indeed the specific sensitivity of GOGAT towards nitric oxide treatment that causes the observed inactivation of glutamate synthase *in vivo*.

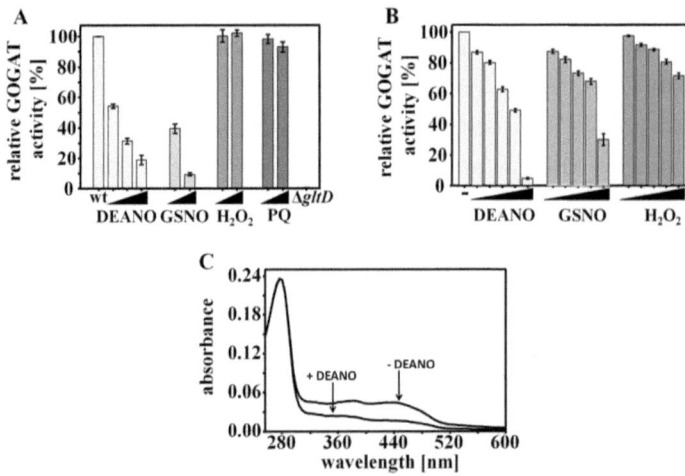

Figure 2.5. Glutamate synthase (GOGAT) is specifically sensitive to nitric oxide treatment. *(A)* E. coli MG1655 *wild type cells were grown in MOPS minimal medium to OD_{600} of 0.4 and treated with increasing concentrations of DEA/NO (0.01 mM, 0.1 mM, 0.5 mM), GSNO (1 mM, 10 mM), H_2O_2 (0.2 mM, 2 mM) or paraquat (0.5 mM, 5 mM). After 10 min of incubation, aliquots were removed and tested for GOGAT activity. GOGAT activity of untreated cells was set to 100%. No GOGAT activity was detected in the ΔgltD deletion strain. Values are mean values from three independent measurements.* **(B)** *Purified GOGAT (200 nM) in 20 mM PPB was treated with increasing concentrations (0.01 mM, 0.1 mM, 0.5 mM, 1 mM, 5 mM) of DEA/NO, GSNO or H_2O_2 for 10 min at 30°C. Then, GOGAT was diluted 1:20 into the GOGAT activity assay. GOGAT activity in the absence of any treatment was set to 100%. The mean and standard deviation of three independent experiments is shown.* **(C)** *Purified GOGAT (4 µM) in 20 mM PPB was either (trace a) left untreated or (trace b) was treated with 1.5 mM DEA/NO for 15 min at 30°C. After removal of DEA/NO using desalting columns, the absorbance spectra were recorded and normalized for the absorption at 280 nm.*

2.3.6 Nitrosative stress treatment causes [4Fe-4S] cluster disassembly of GOGAT *in vitro*

To investigate the influence of nitrosative stress treatment on GOGAT activity in more detail, we purified the wild type protein after overexpression from *E. coli* cells. We then incubated the purified protein in various concentrations of DEA/NO, GSNO or H_2O_2 and analyzed the *in vitro* GOGAT activity. Like *in vivo*, we observed that increasing concentrations of DEA/NO or GSNO cause increasing inactivation of GOGAT while H_2O_2 treatment had small effects on GOGAT activity only at very high concentrations (Fig. 2.5B). These results nicely replicated our *in vivo* data and showed that GOGAT is particularly sensitive to nitrosative stress treatment. Interestingly, however, we did notice that the *in vitro* treatment of GOGAT with DEA/NO was substantially less effective than the *in vivo* treatment (Fig. 2.5A). While 0.01 mM DEA/NO was sufficient to reduce the *in vivo* activity of GOGAT to about 50%, 1 mM DEA/NO was necessary to achieve the same effects *in vitro* (Fig. 2.5B). This decreased efficiency in the reactivity of DEA/NO *in vitro* is likely to be due to the *in vivo* conversion of NO into more thiol-reactive species [217].

To investigate the effects of DEA/NO treatment on the Fe-S clusters in GOGAT, we then analyzed the UV/VIS spectra of GOGAT *in vitro* upon 15 min of treatment in 1.5 mM DEA/NO. This experiment should allow us to monitor the stability and possible nitrosative modifications of GOGAT's [Fe-S] clusters in response to nitric oxide treatment. Purified GOGAT showed the previously described three peaks at 278 nm, 380 nm and 431 nm as well as a broad shoulder at 480 nm (Fig. 2.5B), which is typical for iron-sulfur cluster containing proteins [212]. These signals are contributed by the two [4Fe-4S] clusters of the small subunit and the one [3F-4S] cluster of the large subunit [218]. Upon incubation of GOGAT in 1.5 mM DEA/NO and removal of any residual DEA/NO, a significant decrease over the complete visible absorption range of the [Fe-S] clusters was observed (Fig. 2.5C). In contrast to the redox-sensitive [4Fe-4S] cluster of FNR, where NO-treatment causes the formation of dinitrosyl-iron-cysteine complexes (DNIC) that absorb light around 360 nm [169], no apparent increase in absorption between 300-360 nm was observed in our sample. This result suggested that at least in our *in vitro* experiments, NO treatment causes the disassembly of GOGAT's [Fe-S] clusters possibly by modifying one or more of the cysteines involved in cluster assembly. Similar effects of RNS treatment on [4Fe-4S]$^{2+}$ clusters has to our knowledge only been observed in aconitase, where peroxynitrite treatment was found to cause cluster disassembly by oxidizing one of the cluster-coordinating cysteines [219].

2.3.7 RNS targets cysteines of GOGAT's [Fe-S] cluster

To investigate whether RNS-mediated cysteine modifications are indeed responsible for the observed [Fe-S] cluster disassembly, we determined the redox status of the individual cysteines in DEA/NO-treated GOGAT using mass spectrometry. To visualize and compare the redox status of the individual cysteines in untreated and DEA/NO-oxidized GOGAT, we incubated both proteins with the alkylating agent N-ethylmaleimide (NEM) under denaturing conditions. NEM irreversibly binds to all reduced thiols and leads to the mass addition of 125 Da per cysteine. In the next step, all reversible thiol modifications such as disulfide bonds or sulfenic acids present in the GOGAT preparations were reduced with DTT. These newly exposed thiol groups were then alkylated with the thiol reactive reagent iodoacetamide, which adds 52 Da per cysteine. After this differential labeling process, proteins were digested with trypsin and peptides were analyzed by mass spectrometry. Based on this trapping protocol, peptides that harbor originally reduced cysteines will reveal their peptide mass including a 125 Da mass addition per cysteine, while the same peptide with originally oxidized cysteines will only have a 52 Da mass addition per cysteine (Fig. 2.6). Although this method is not quantitative, it allows us to directly monitor the relative changes in the distribution of reduced and oxidized peptides.

We found five of GOGAT's 11 cysteines to be heavily oxidatively modified upon DEA/NO treatment *in vitro* (Fig. 2.6). Cysteines Cys94, Cys104, Cys108 are known to be involved in the formation of one of the two [Fe-S] clusters in GOGAT [220]. Oxidative modification of these cysteines supports our spectroscopic studies that show the disassembly of GOGAT's [Fe-S] cluster(s) (Fig. 2.5C). Our finding that substantial reactivation of GOGAT can be observed upon recovery of *E. coli* cells from DEA/NO treatment in the presence of chloramphenicol (data not shown) suggests, however, that GOGAT's [Fe-S] clusters can be repaired in aerobically growing *E. coli* cells. In addition to the three cysteines that are involved in Fe-S cluster formation, we also found two of the three non-cluster associated cysteines of GOGAT (Cys161, Cys302) to be significantly oxidatively modified (Fig. 2.6). All other cysteines were either found to be DEA/NO-resistant (Cys59) or were not identified in our mass spectrometric data. These results indicate that RNS-mediated thiol modifications lead to [Fe-S] cluster disassembly in GOGAT. It remains to be determined whether the cysteines that coordinate the [Fe-S] cluster(s) are the primary RNS-targets or whether they only become targets once the [Fe-S] cluster has disassembled.

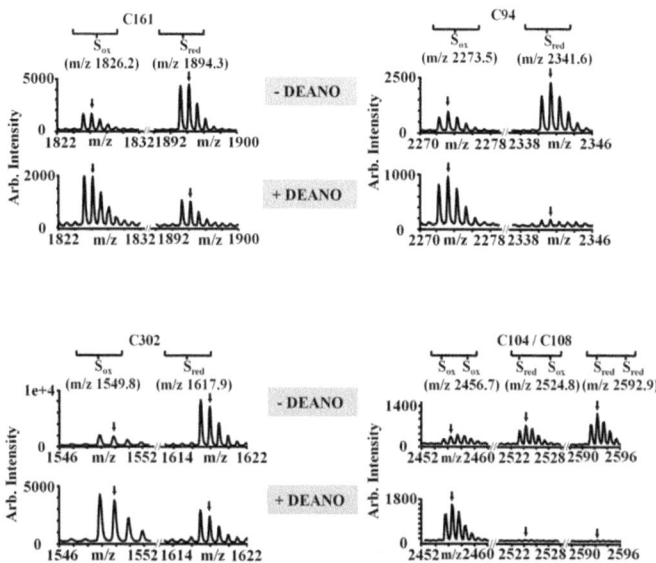

Figure 2.6. Analysis of GOGAT's thiol status using mass spectrometry. *Purified GOGAT (4 µM) in 20 mM PPB buffer that was either left untreated or oxidized with 0.5 mM DEA/NO for 15 min at 30°C were first incubated with N-ethylmaleimide (125 Da per cysteine), followed by DTT reduction and iodoacetamide (57 Da per cysteine) treatment. Mass spectra of the differentially labeled tryptic peptides were obtained by MALDI-MS and specific regions of these spectra are presented for the tryptic peptides, which were found to harbor significantly oxidized cysteines upon DEA/NO treatment. Cys161 and Cys302 are both non-[Fe-S] cluster associated cysteines, while Cys94, Cys104 and Cys108 coordinate one of the two Fe-S clusters in GOGAT. For Cys161 (aa 149-167), m/z value of 1894.3 represents fully reduced peptide while m/z value of 1826.2 corresponds to fully oxidized peptide. For Cys302 (aa 290-304), m/z value of 1617.9 represents the fully reduced peptide while m/z value of 1549.8 corresponds to the fully oxidized peptide. For Cys94 (aa 77-96), m/z value of 2341.6 represents the fully reduced peptide, m/z value of 2273.5 corresponds to the fully oxidized peptide. For Cys104/Cys108 (aa 103-124), m/z of 2592.9 represents the fully reduced peptide while m/z of 2456.7 corresponds to the fully oxidized peptide and m/z of 2524.8 represents the peptide harboring one NEM and one IAM label. Arrows mark expected theoretical mass of the modified peptide.*

2.4 Conclusions

We show here that sublethal concentrations of nitric oxide cause the oxidative modification of a specific subset of thiol-containing *E. coli* proteins, many of which have been previously shown to be encoded by genes that are essential under our growth conditions [221]. Media supplementation studies provided evidence that the growth inhibitory effect of nitric oxide is due to oxidative modification of enzymes of the pyruvate dehydrogenase complex as well as enzymes of the branched chain amino acid biosynthesis pathway. While the nitrosative stress conditions that *E. coli* cells experience in our laboratory setting are short-term and the effects are largely reversible, bacteria that become targets of the mammalian host defense presumably experience high concentrations of reactive nitrogen species over longer periods of time [222]. We assume that, at a minimum, the same set of bacterial proteins will be affected in these cells. Long term inactivation of these essential proteins, potential overoxidation of the affected cysteines, or additional non-thiol modifications might then lead to irreversible protein damage and to the lethal effects of reactive nitrogen species.

Interestingly, our RNS-sensitive proteins are largely distinct from the set of proteins that we previously identified to harbor H_2O_2-sensitive cysteines. This specificity appears to make physiological sense because reactive oxygen and nitrogen species differ in their chemistry and require distinct cellular responses and defense systems [42,222,223]. It will be our future challenge to determine what makes these proteins particularly sensitive to oxidative and nitrosative stressors and distinguishes them from the hundreds of cysteine-containing *E. coli* proteins, whose thiol groups are not affected by ROS and RNS. Specific nitrosylation motifs have been suggested [224], however, only one of our identified proteins (Ef-Tu) was found to harbor a cysteine in such motif and many abundant proteins that harbor these motifs have not been identified in our study (data not shown). It is therefore very likely that amino acids or prosthetic groups such as [Fe-S] clusters, which are nearby in the tertiary structure, are even more important in determining the reactivity of individual cysteine residues.

This conclusion also agrees with earlier studies that showed that the structural microenvironment of the cysteine S_γ-atom is absolutely critical [224]. Therefore, unless we have the three-dimensional structure of many of these NO- and H_2O_2-sensitive proteins, it will be very difficult to predict redox-regulated proteins simply based on their amino acid sequence. We can, however, exploit their specific oxidative and nitrosative stress sensitivity for diagnostic purposes by determining marker proteins that are specifically sensitive to distinct reactive oxygen and nitrogen species. Identification of such marker proteins in cells and tissues can thus be used as direct read-out for the presence of specific reactive oxygen or nitrogen species.

3 Oxidation-sensitive yeast proteins in subcellular compartments

Chapter 4 contains the manuscript "Oxidation-sensitive yeast proteins in subcellular compartments"; Brandes, N., A. Lindemann, and U. Jakob, 2009, in preparation for publication

Abstract

Oxidative thiol modifications influence the activity and structure of numerous proteins, which are involved in many different cellular processes. To understand the effects of oxidative stress on cells and organisms, it is thus crucial to identify redox-sensitive proteins, the targets of oxidative stress *in vivo*. We used a global proteomic approach to precisely quantify the *in vivo* redox status of almost 300 yeast proteins located in the cytosol and subcellular compartments (*i.e.*, mitochondria, nuclei, vacuoles). To identify targets of specific reactive oxygen species, we treated yeast cells with sublethal concentrations of H_2O_2 and the superoxide-generating agent paraquat. We found 60 proteins that were specifically sensitive towards H_2O_2 treatment, four proteins that revealed thiol modifications in response to superoxide treatment and four proteins that were sensitive towards both oxidants. Our results indicate that H_2O_2 affects preferably key metabolic enzymes involved in translation, carbohydrate metabolic processes and amino acid biosynthesis. Paraquat-induced superoxide generation causes oxidative thiol modifications especially to proteins of the amino acid biosynthesis and the TCA cycle enzyme aconitase. These targets are primarily located in the mitochondria, the central superoxide-producing organelle during respiratory growth. Our results reveal that the quantitative assessment of a protein's oxidation state is a valuable tool to identify catalytically active and redox-sensitive cysteine residues and illustrate how cells use reversible oxidative thiol modifications to coordinate a metabolic response, which likely contributes to the protection against the effects of oxidative stress.

3.1 Introduction

The intracellular accumulation of reactive oxygen species (ROS) causes a cellular condition generally termed oxidative stress. Oxidative stress is attributed to numerous physiological and pathological conditions such as cancer, neurodegenerative diseases, diabetes [22,23] and is thought to be the underlying culprit of eukaryotic aging [5]. Inside the cells, ROS like hydrogen peroxide (H_2O_2), superoxide anions ($\cdot O_2^-$), and hydroxyl radicals ($\cdot OH$) are constantly generated as byproducts of respiration, products of NADPH oxidases and in peroxisomes [3,4]. Cells use a number of enzymatic and non-enzymatic defense and repair strategies to preserve an overall reducing redox environment while, at the same time, allowing for the presence of low ROS concentrations, which are involved in physiological signaling processes [5,6,12,13]. Once ROS concentrations exceed the antioxidant capacity of the cell, however, oxidative damage to DNA, lipids, and proteins can occur (Fig. 3.1) [8,18,19].

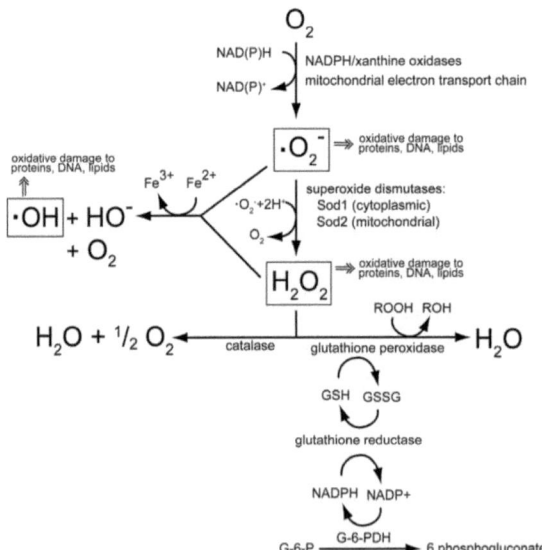

Figure 3.1. Major pathways of reactive oxygen species (ROS) generation and detoxification. *Superoxide anion radicals ($\cdot O_2^-$) are generated as products of xanthine and NADPH oxidases, or as byproducts of the mitochondrial electron transport chain. Superoxide dismutases (SODs) convert superoxide into hydrogen peroxide (H_2O_2), which is converted to H_2O and molecular oxygen by catalases or peroxidases, such as the glutathione peroxidases. Glutathione peroxidases use reduced glutathione (GSH) as electron donor, which is oxidized to GSSG in this process. Glutathione reductase then recycles GSSG in a NADPH-consuming step back to GSH, catalyzed by glucose-6-phosphate dehydrogenase (G-6-PDH). The most toxic ROS appears to be hydroxyl*

radicals (•OH), formed by Fenton chemistry when •O_2^- and H_2O_2 are exposed to trace transition metals, such as Fe^{2+} or Cu^+.

Over the last few years, it has been shown that an increasing number of proteins are not damaged by ROS but use ROS-mediated thiol modifications to regulate their function, structure or localization. In this process, the reactive thiol groups of cysteines undergo a variety of different oxidative modifications. *In vivo* reversible oxidative thiol modifications include sulfenic (R-SOH) acid and disulfide bond formation, S-nitrosylation or S-glutathionylation. Irreversible reactions involve oxidation to sulfinic (R-SO_2H) or sulfonic (R-SO_3H) acids.

Many redox-sensitive proteins are involved in important cellular processes such as transcription, metabolism, stress protection, apoptosis, and signal transduction [38,70,225]. Redox-regulated proteins not only act in the cytoplasm but also in subcellular compartments of the eukaryotic cell, including mitochondria, nuclei, vacuoles, ER, and peroxisomes [56,77,226-228]. Knowledge of the precise redox status of these oxidation-sensitive proteins is therefore crucial for our understanding of a cell's physiology. For instance, the majority of ROS is thought to be generated in mitochondria as side-products of oxidative phosphorylation [229,230]. As many enzymes involved in oxidative phosphorylation contain redox-sensitive cysteine residues themselves, the mitochondrial proteome is therefore likely to be especially susceptible to oxidative stress and damage [226,227]. That maintaining the mitochondrial redox homeostasis is critically important is further supported by findings that mitochondria-based thiol modifications are linked to many oxidative stress-induced diseases such as Parkinson [231], Friedreich's ataxia [232], Huntington [233], and diabetes [234]. Similarly, nuclei harbor a multitude of important redox-regulated cell processes that are controlled by cysteine-containing proteins such as DNA synthesis [228], regulation of transcription factors [56,77] and nuclear matrix organization [235].

Analysis of the redox status of proteins in subcellular compartments has been compounded by the fact that purification of organelles, albeit necessary to detect low-abundance proteins, induces changes in the redox status of the proteins [54,236]. To circumvent these problems, redox western blots were performed to identify the redox status of defined, selected proteins [237,238]. In addition, expression of redox-sensitive variants of green fluorescent protein (GFP), FRET-measurements and determination of subcellular GSH:GSSG ratios [237,239,240] have been used to obtain the redox potentials of major subcellular compartments. These measurements showed that the redox potential arranged from most reducing to most oxidizing follows the order: cytosol alone (~ -290 mV) [239] < isolated mitochondria (-250 mV to -280 mV) [241] < whole cells (-221 mV to -236 mV) [242] < ER (-170 mV to -185 mV) [242] < extracellular space [243]. The redox potential of nuclei is suggested to be similar to the cytoplasm as GSH and GSSG can readily diffuse through the nuclear envelope [244]. While these mostly gel-based proteomic approaches were designed to determine the redox state of individual proteins, they did not provide a global quantitative assessment of the cellular redox state. Additionally, many of these methods lacked the ability to identify oxidation-sensitive cysteines in redox-regulated cysteines involved [51,237,245,246].

With the awareness that oxidative thiol modifications of redox-sensitive proteins often influence structure and activity of the proteins, it is crucial to assess the quantitative extent of reversible oxidation of proteins and their affected cysteine residues *in vivo*. This knowledge should help us to i) identify proteins especially prone to oxidation, ii) to determine proteins that are potentially involved in cellular antioxidant defense response and iii) to find proteins that are targets of specific ROS, which could serve as intracellular marker proteins. Thus, we set out to globally quantify the *in vivo* thiol status of yeast proteins in the cytosol and sub-cellular compartments (*i.e.*, mitochondria, nuclei, vacuoles) of exponentially growing yeast cells by using our recently established mass spectrometry based thiol trapping technique termed OxICAT [49]. We determined the steady-state redox level of almost 300 yeast proteins. In agreement with the existing literature, the majority of proteins were found to be predominantly reduced. Only few of the identified proteins were fully oxidized, with most of these protein's cysteines fulfilling structural functions. Treatment of yeast cells with H_2O_2 or the superoxide-generating agent paraquat revealed that 60 proteins were specifically sensitive towards H_2O_2, four proteins that were oxidized by superoxide and four proteins that were affected by both oxidants. Enzymes that fulfill important roles in translation, carbohydrate metabolic processes and amino acid biosynthesis appear to be the main cellular targets of peroxide stress. Superoxide preferably affects proteins of the amino acid biosynthesis and aconitase, a key player in the TCA cycle. These results suggest that cells use posttranslational thiol modifications to ensure the continuous production of the reducing equivalent NADPH and stop wasteful energy consumption during times of peroxide- and superoxide-mediated stress.

3.2 Material and Methods

3.2.1 Strains and cell growth

For all OxICAT experiments, the S. cerevisiae strain EG103 (DBY746; MATα, leu2-3 112 his3Δ1 trp1-289a ura3-52) was grown aerobically at 30°C in synthetic complete (SC) medium (0.67% yeast nitrogen base without amino acids) supplemented with complete amino acid dropout solution and 2% glucose (synthetic complete dextrose; herein referred to as "SCD media") or 3% glycerol (synthetic complete glycerol; herein referred to as "SCG media") as carbon source. Cells were picked from fresh colonies and grown twice to late logarithmic phase in SCD or SCG media. Cells were then diluted into fresh media using flasks with volume-to-medium ratio of 5:1. Growth was monitored by measuring the optical density at OD_{600}. For organelle purification, cells were either grown in YPD (1% (w/v) yeast extract, 2% (w/v) bactopeptone, 2% (w/v) glucose) or YPG (1% (w/v) yeast extract, 2% (w/v) bactopeptone, 3% (w/v) glycerol, pH 5.0) medium.

3.2.2 Analysis of glucose concentration

Glucose concentration in the growth media was analyzed using an enzymatic assay kit (Glucose (HK) Assay Kit, Sigma-Aldrich) that combines hexokinase and glucose-6-phosphate dehydrogenase activities [247]. For the measurements, 10 – 200 µl of the growth media (~ 0.5 – 50 µg of glucose) was added to 1 ml of "Glucose assay reagent" (prepared according to manufacturer's protocol), mixed and incubated for 15 min at 25°C. Then, the absorbance was determined at 340 nm using a Cary 3 UV spectrophotometer (Varian Instruments). The glucose concentration was calculated according to the manufacturer's protocol. Blanks were prepared by adding water to the assay reagent instead of glucose-containing growth media.

3.2.3 Purification of subcellular compartments

3.2.3.1 Purification of mitochondria

The purification of mitochondria is based on Meisinger et al. [248]. To increase the mitochondria yield, yeast strain S. cerevisiae EG103 was grown in YPG media at 30°C with vigorous shaking until an OD_{600} of 2. Then, cells were harvested by centrifugation (3,000 x g, 5 min), washed twice with $_{dd}H_2O$, resuspended in 2 ml/g wet weight resuspension buffer (100 mM Tris-H_2SO_4 pH 9.4, 10 mM DTT) and incubated for 20 min at 30°C and 30 rpm. After being washed with 7 ml/g zymolyase buffer (1.2 M sorbitol, 20 mM potassium phosphate pH 7.4), cells were incubated with 7 mg per gram wet weight Zymolyase-20T (Seikagaku Kogyo Co.) in 7 ml per gram wet weight zymolyase buffer with 30 rpm for 45 min at 30°C for conversion into spheroplasts. Cells were harvested (1,500 x *g*, 5 min, 4°C) and washed twice with zymolyase buffer. From this step onwards, the lysed material and the equipment was maintained at low temperature to avoid proteolysis. The pellet was resuspended in 6.5 ml per gram wet weight ice-cold homogenization buffer (0.6 M sorbitol, 10 mM Tris-HCl pH 7.4, 1 mM EDTA, 1 mM PMSF, 0.2% (w/v) BSA) and homogenized with 15 strokes in a glass-teflon homogenizer. This homogenate was then diluted with 1 vol of ice-cold homogenization buffer and cell debris and nuclei were removed by centrifugation (1,500 x *g*, 5 min). The supernatant was centrifuged (3,000 x *g*, 5 min) and mitochondria were pelleted from the supernatant (12,000 x *g*, 15 min). The mitochondrial pellet was washed with SEM buffer (250 mM sucrose, 1 mM EDTA, 10 mM Mops pH 7.2) and pelleted again (12,000 x *g*, 15 min). The resulting crude mitochondrial fraction was resuspended in SEM buffer to a concentration of 10 mg/ml.

To remove further contaminations, the crude mitochondrial fraction was treated with 10 strokes in a glass-teflon homogenizer and immediately loaded onto a sucrose step gradient of 1.5 ml 60%, 4 ml 32%, 1.5 ml 23%, 1.5 ml 15% (w/v) sucrose in EM buffer (10 mM Mops pH 7.2, 1 mM EDTA). After centrifugation in a SW41 Ti rotor (134,000 x *g*, 1 h, 4°C), the purified mitochondria were recovered from the 60%/32% interface. The mitochondria were then diluted with 2 vol of SEM buffer, pelleted (10,000 x *g*, 10 min), adjusted with SEM to 10 mg/ml, shock-frozen with liquid N_2 and stored at -80°C. Western blot analysis with purified mitochondria and the mitochondrial markers Tom40 and Porin, the vacuolar marker Alp, the cytoplasmic marker Pgk1, and the ER markers Sec61 and Sss1 showed an enrichment of mitochondrial proteins and the depletion of contaminants.

3.2.3.2 Purification of vacuoles

The purification of vacuoles is based on Haas [249]. *S. cerevisiae EG103* was grown in YPD media at 30°C with vigorous shaking until late logarithmic phase and spheroplast formation was conducted as described for mitochondria purification. After washing of spheroplasts with zymolyase buffer, cells were harvested (1,500 x *g*, 5 min, 4°C) and 1.5 ml per gram wet weight of pre-chilled 15% (w/v) Ficoll solution in Ficoll gradient buffer (10 mM PIPES/KOH pH 6.8, 200 mM sorbitol, 1 mM PMSF, 1 x protease inhibitor cocktail (PIC)) was added to the pellet. Then, 150 µl of 0.4 mg/ml dextran solution in 15% Ficoll was added and spheroplasts were incubated for 2 min on ice. Cell membranes were disrupted by incubation for 60 sec at 30°C and the cell lysate was immediately placed on ice. After transferring the lysate into ultracentrifugation tubes, 2 ml per gram wet weight 8% (w/v) Ficoll solution and 2.5 ml per gram wet weight 4% (w/v) Ficoll solution were carefully layered over the cell lysate and tubes were filled with Ficoll buffer. Centrifugation was conducted in a SW41 Ti rotor (110,000 x g, 90 min, 4°C). The vacuolar fraction was removed from the Ficoll buffer/4% Ficoll interface using a trimmed 200 µl pipette tip. Purity of organelles was monitored by phase-contrast microscopy and protein concentration was determined using Bradford. Vacuoles were shock-frozen with liquid N_2 and stored at -20°C.

3.2.3.3 Purification of nuclei

The purification of nuclei is based on Dove et al. [250]. *S. cerevisiae EG103* was grown in YPD media at 30°C with vigorous shaking until late logarithmic phase and spheroplast formation was conducted as described for mitochondria purification. After washing, the pellet was resuspended in 4 ml per gram wet weight Zymolyase buffer. 1 ml per gram wet weight of Ficoll cushion solution (20% (w/v) sorbitol, 5% (w/v) Ficoll 400, 1x PIC) was layered under the spheroplast suspension and spheroplasts were centrifuged through the cushion solution (4,000 x g, 10 min, 4°C). 5 ml per gram wet weight of 20% (w/v) Ficoll solution was added to the pellet and the cells were immediately lysed in a pre-chilled homogenizer with 20 strokes. After incubation on ice for 15 min, the lysate was centrifuged twice (13,000 x g, 10 min, 4°C) and the supernatant was layered onto a pre-chilled Ficoll step gradient (6.5 g of 50%, 40%, and 30% (w/v) Ficoll solution each, 1x of PIC). The Ficoll step gradient was centrifuged (58,400 x g, 60 min, 4°C) and nuclei were collected between the 30%/40% and 40%/50% Ficoll interfaces and throughout the 40% Ficoll layer (8 ml total volume). Then, the nuclei/Ficoll suspension was diluted with 1 vol of cold 1x PM buffer (5 mM K_2HPO_4, 15 mM KH_2PO_4, 1 mM $MgCl_2$, 1x PIC) and harvested nuclei were again layered over a second pre-chilled Ficoll step gradient (5.5 g of 50%, 40%, and 30% (w/v) Ficoll solution each, 1x of PIC). The gradient was centrifuged (58,400 x g, 60 min, 4°C) and 10 ml were harvested as described above. To remove residual Ficoll solution, nuclei were diluted in 10 vol cold 1x PM buffer and pelleted by centrifugation (10,000 x g, 10 min, 4°C). Protein concentration was determined using Bradford. Vacuoles were shock-frozen with liquid N_2 and stored at -20°C.

3.2.4 Differential thiol trapping of whole cells using ICAT

3.2.4.1 Differential thiol trapping procedure

Yeast cells were grown in SCD medium at 30°C with continuous shaking until midlogarithmic phase was reached. For oxidant treatment, cultures were split and incubated with 0.5 mM hydrogen peroxide (H_2O_2) or 3 mM paraquat (superoxide-generating agent) for 15 min. 12 ml of cell culture (~100 µg of protein) were harvested either before oxidant addition (used to evaluate steady-state oxidation levels) or 15 min after oxidant treatment onto trichloroacetic acid (TCA) (10% (w/v) final concentration) to stop all thiol-disulfide exchange reactions and incubated on ice for 30 min. Then, the thiol trapping protocol established by Leichert et al. [49] was followed with few modifications to adjust for yeast cell lysis. TCA precipitates were centrifuged (13,000 x g, 30 min, 4°C) and the pellet was washed with 500 µl of ice cold 10% (w/v) TCA and 200 µl of ice cold 5% (w/v) TCA. To prevent any air oxygen-mediated oxidative thiol modifications in redox-sensitive S. cerevisiae proteins during the thiol trapping procedure, the labeling step with light ICAT was conducted inside an anaerobic chamber. The pellet was dissolved in 80 µl of denaturing alkylation buffer (DAB; 6 M urea, 0.5% (w/v) sodium dodecyl sulfate (SDS), 10 mM EDTA, 200 mM Tris-HCl pH 8.5) and the contents of one vial of cleavable light ICAT reagent (Applied Biosystems, Foster City, CA) dissolved in 20 µl of acetonitrile (ACN). The cells were disrupted in the presence of DAB/light ICAT using glass-beads and rigorous vortexing for 4 x 45 seconds. The supernatant was separated from glass-beads and cell debris by centrifugation (13,000 x g, 2 min), and incubated at 1,300 rpm for 2 h at 37°C in the dark. To remove light ICAT, the proteins were precipitated with 500 µl of prechilled (–20°C) acetone for > 4 h at –20°C. After centrifugation (13,000 x g, 30 min, 4°C), the pellet was washed twice with 500 µl of pre-chilled acetone. Then, the protein pellet was dissolved in a mixture of 80 µl of DAB, 2 µl of 50 mM Tris(2-carboxyethyl)phosphine hydrochloride (TCEP), and the contents of one vial of cleavable heavy ICAT dissolved in 20 µl of ACN. The sample was incubated at 1,300 rpm for 2 h at 37°C in the dark. Proteins were then acetone-precipitated and washed as described above.

3.2.4.2 Tryptic digest and purification of ICAT-labeled peptides

The TCA pellet of the ICAT labeled proteins was re-dissolved in 80 µl of 0.1% SDS, 50 mM Tris-HCl pH 8.5, 20 µl of acetonitrile, and 100 µl of L-1-tosylamido-2-phenylethyl chloromethyl ketone (TPCK) treated trypsin solution (25 µg diluted in 200 µl) (Applied

Biosystems). All subsequent steps were performed according to the protocol provided by Applied Biosystems. After the purification of the peptides using avidin column chromatography, the eluate was evaporated to dryness in a vacuum centrifuge and re-dissolved in 5% (v/v) triisopropylsilane in trifluoroacetic acid. The samples were incubated for 2 h at 37°C to cleave the biotin moiety from the ICAT residue. Subsequently, the samples were evaporated to dryness in a vacuum centrifuge for LC-MS/MS analysis.

3.2.5 LC-MS/MS analysis

LC-MS/MS analysis was conducted by the Michigan Proteome Consortium at the University of Michigan. Liquid chromatography (LC) was performed using an 1100 Series nano HPLC system equipped with μWPS autosampler, 2/10 microvalve, MWD UV detector (at 214 nm) and Micro-FC fraction collector/spotter. Peptides were reconstituted with 0.1% (v/v) TFA and then enriched and desalted on a 5 x 0.3 mm Zorbax 300 SB-C18 Column with 5 μm particle size (Agilent technologies, Palo Alto, CA). The loading and washing solution was 3% (v/v) Acetonitrile (ACN), 0.1% (v/v) trifluoroacetic acid (TFA), flowing at 20 μl/min for 9 min. This pre-column was then attached in front of a 100 x 0.1 mm 300 SB-C18 column with 3.5 μm particle size (Micro-Tech Scientific, Vista, CA) and the system equilibrated with solvent A (Water:TFA, 99.9:0.1). Peptides were eluted with a gradient of solvent B (ACN:Water:TFA, 90:10:0.1) from 6.5% B to 50% B over 90 min at a flow rate of 0.4 μl/min. Column effluent was mixed with matrix (2 mg/ml α-CHCA in Methanol:Isopropanol:ACN:Water:Acetic acid (12:33.3:52:36:0.7) containing 10 mM ammonium phosphate) delivered with a PHD200 infusion pump (Harvard Apparatus, Hamden, CT) at 1 μl/min and fractions were directly spotted at 28 second intervals onto stainless steel maldi targets.

Mass spectra for each fraction were acquired on an Applied Biosystems 4800 Maldi TOF/TOF Analyzer. Laser settings for the parental spectrum were set to 1500 shots, using a 200 Hz YAG laser operated in the 3rd harmonic (355 nm). The TOF/TOF was operated in positive ion reflectron mode. Seven point Gaussian smoothing was applied to spectra and a signal to noise filter (S/N=30) applied for peak picking. Eight wells on the MALDI target plate were calibrated with the peptide standards [Glu]^1 fibrinopeptide B (m/z 1570.677), ACTH (m/z 2465.199), angiotensin (m/z 1296.685), bradykinin (m/z 904.468), and ACTH 1-17 (m/z 2093.087) (all m/z values denote the mass of the unlabeled peptide). These individually calibrated wells were used to update the default MS calibration. A full plate calibration correction was calculated based on those eight wells and was applied to the default calibration for both MS and MS/MS spectra. Initially, the eight most intense peaks in each MS spectrum were selected for MS/MS analysis. If particular features of interest were later found in the OxICAT data analysis, additional MS/MS spectra were acquired from the plate with manually selected precursors. MS/MS spectra were obtained from a maximum of 4000 laser shots, depending on spectra quality. Fragmentation of the labeled peptides was induced by the use of ambient air as a collision gas with a pressure of ~6 x 10^{-7} torr and a collision energy of 2 kV.

Peptide identifications were performed using GPS Explorer software (Applied Biosystems), which acts as a front end for the Mascot search engine (MatrixScience, London, UK). Each MS/MS spectrum was searched against Swiss-Prot filtering on *S. cerevisiae*. As parameters, trypsin specificity with one missed cleavage and ^{12}C-ICAT and ^{13}C-ICAT as variable cysteine modifications was used. Oxidized methionine and N-terminal pyroglutamyl formation was considered as an additional variable modification. The precursor tolerance and MS/MS fragment tolerances were set to ± 0.7 and ± 0.3 Da, respectively. Identification was considered significant if a "Total Ion Score C.I. %" value of 95 or above was assigned by the software.

3.2.6 Data analysis

OxICAT data were analyzed by using an extension of the existing open-source program *msInspect* [251]. Mass signals with identical elution profiles and a mass difference of 9 Da or multiples thereof were combined to generate "ICAT pairs". If an ICAT pair was found in more than one fraction, the ICAT pair in the fraction with the highest intensity was used for the calculation of the percentage of oxidized peptide. For the calculation of the percentage of oxidation of any given peptide, the average percentage of oxidized peptide of at least three independent experiments was used and the results were expressed as mean ± standard deviation. Peptides that revealed an increase of at least 1.4-fold in the fraction of oxidized peptide upon oxidative stress treatment when compared with control conditions were considered as potentially redox-sensitive.

3.2.7 Design of the OxICAT program by extending msInspect

The open source program *msInspect* is designed to display and quantify mass spectrometry data derived from liquid chromatography coupled MS runs [251]. The workflow for the OxICAT data analysis was performed in the following six steps:

3.2.7.1 File conversion

The mass spectrometry data were exported as mass/intensity tab delimited text files from the spectra obtained by the 4000 Series Explorer software. They were then converted to mzXML files using the program ConvertMzXML (downloadable at tranche.proteomecommons.org).

3.2.7.2 Detection of peptides

Peptides (*i.e.*, features) are represented as isotope peak clusters in the mass spectrum. In peptide detection, the monoisotopic peak in each peak cluster is assigned to a unique peptide identifier. Peak detection was performed for every spectrum using the 4000 Series Explorer software from Applied Biosystems resulting in a peak list. Then, a baseline and noise-line was calculated for each spectrum by smoothing the lowest and the highest intensity values per Da over a window of 20 Da. Detected peaks were rejected if the signal-to-noise ratio was found to be below a defined threshold of 4:1. Most peptides eluted over several fractions. The detected peaks in all peak files (scan) were combined according to their monoisotopic mass, and assigned a unique peptide ID if they eluted over consecutive fractions. A gap parameter was introduced, which allows for interruption of the series to account for low quality scans.

3.2.7.3 Retention time correction and peptide matching

To facilitate the comparison of detected peptides between samples, a linear retention time correction was performed. Every peptide detected in one sample was then matched to the corresponding peptide in another samples allowing for a variance in both mass and scan (0.2 Da, 2 scans). In some instances, no matching peptide was found, either because of a failure in the peak detection algorithm or, more likely, because of the true absence of a relevant signal. To narrow the search space in the sample in which the respective peptide was not detectable, the program was designed to average both the m/z values and scans of the peptide from all the corresponding samples, where the feature was detected. These coordinates were then used as starting point to find the closest local maximum, which was quantified and matched.

3.2.7.4 Detection of ICAT pairs

Peptides, which differ by multiples of 9 Da on the mass scale and elute in the same fraction were combined to OxICAT multiplets and assigned a unique identifier. Like before, this process was designed to allow for a certain m/z as well as scan shift. Each OxICAT multiplet contains the different masses representing the different oxidation states of a single peptide. Steps 3.2.7.2 to 3.2.7.4 were performed using the OxICAT extension written for *msInspect* [49,251].

3.2.7.5 Quantification of the oxidation state

The next step was to determine the status of oxidation for all OxICAT multiplets. The signal with the lowest *m/z* value in an OxICAT multiplet was assumed to correspond to the fully reduced state of the peptide. The signal with the highest *m/z* value was assumed to represent the fully oxidized form. If the peptide contained more than one cysteine, mixed intermediate forms could exist as well. The computed percentage oxidized values reflect the portion of each peptide being oxidized, including partially oxidized intermediates. With the help of the peptide matching, the change of the oxidation state of peptides under different conditions was determined. Step 3.2.7.5 was performed using a script that analyzed the "*correctedICAT.tsv*" result file from the *msInspect* OxICAT extension. The script is available at tranche.proteomecommons.org.

3.2.7.6 Quality control

For a peptide to be considered redox-sensitive, several requirements had to be fulfilled. The peptide had to be on average at least 1.4-fold more oxidized under the chosen oxidative stress conditions as compared to control conditions. Moreover, the largest peak in an OxICAT multiplet had to be at least 3-fold above background in the respective sample. These calculations were performed with Excel spreadsheet calculation Software (Microsoft, Redmond, WA). All redox-regulated peptides were additionally verified using the visualization tools in *msInspect*.

Chapter 3: Oxidation-sensitive *S. cerevisiae* proteins 73

3.2.8 Enhanced detection of organelle-specific yeast proteins using OxICAT spiking

For ICAT labeling of organelle-specific yeast proteins, 200 µg of the respective organelle protein preparation (*i.e.*, nuclei, vacuole, or mitochondria) was added to TCA (10% (w/v) final concentration) and incubated on ice for 30 min. TCA precipitates were centrifuged (13,000 x g, 30 min, 4°C) and washed as described above. The p ellet was dissolved in 160 µl of DAB and 4 µl of 50 mM TCEP, split up into equal volumes and added to the contents of either one vial of cleavable light or heavy ICAT reagent, respectively, dissolved in 20 µl of ACN. The samples were then incubated at 1,300 rpm for 2 h at 37°C in the dark. To remove the ICAT reagent, proteins were precipitated with 500 µl of pre-chilled acetone for >4 h at –20°C. After centrifugation (13,000 x g, 30 min, 4°C), the pellet was washed twice with 500 µl of pre-chilled acetone and tryptic digest of the ICAT-labeled peptides was conducted as described before. After enrichment of peptides on streptavidin columns, light and heavy labeled peptides eluted in "Affinity Buffer Elute" (provided by Applied Biosystems) were combined in a 1:1 ratio and subjected to biotin tag cleavage and LC-MS/MS. For preparation of LC-MS samples, peptides in "Affinity Buffer Elute" of total cell lysate were spiked with heavy and light labeled peptides in a 6:3:1 ratio (Cell lysate:Heavy:Light), and then subjected to biotin tag cleavage and LC-MS.

3.2.9 Prediction of pK_a value, cysteine accessibility, secondary structure and fold propensity

To calculate and predict pK_a values and accessibility of H_2O_2-sensitive and insensitive cysteine residues, the prediction program PROPKA 2.0 (http://propka.ki.ku.dk/) was used [252]. Calculations of cysteine-containing secondary structures were performed by STRIDE (http://molbio.info.nih.gov/ structbio/basic.html) [253], a program which uses hydrogen bond energy and main chain dihedral angles to recognize secondary structural elements in proteins from their atomic coordinates. For both PROPKA and STRIDE, the available crystal structures of 20 H_2O_2-sensitive and 72 insensitive cysteine-containing proteins served as input. If more than one structure per protein was available, the structure with the highest resolution was chosen. Fold predictions to estimate the probability of a cysteine-containing amino acid stretch to fold were performed with FoldIndex (http://bip.weizmann.ac.il/fldbin/findex/) [254]. For FoldIndex, default settings were chosen and the amino acid sequence of 295 H_2O_2-insensitive and 61 H_2O_2-sensitive cysteine-containing proteins served as input.

3.3 Results and Discussion

3.3.1 OxICAT as method to identify the redox state of proteins in yeast

In the past decade, several redox-sensitive cytoplasmic and organelle-specific proteins in yeast have been discovered by proteomic approaches, which contributed to the elucidation of their functional importance in numerous cellular processes [255,256]. However, most proteomic studies that investigated the oxidation state and sensitivity of individual proteins in exponentially growing yeast cells did not pay special attention towards organelle-specific proteins [51,245,257]. Moreover, none of these studies used a quantitative approach to assess the extent of their steady-state redox status and their level of ROS-mediated oxidation. Thus, little is known about the degree of oxidative modification of these proteins during logarithmic growth and their specific sensitivity towards reactive oxygen species. But since the degree of oxidative modification provides us with valuable knowledge about activity, regulation, and potentially the structure of a protein, this information is crucial for understanding not only the function and molecular mechanism of a protein but also organelle-specific processes and interactions with other cell components. Here, we used a modified version of our recently developed thiol trapping technique OxICAT [49] to quantitatively assess the steady-state oxidation of yeast proteins located in the cytoplasm, mitochondria, nuclei and vacuoles in exponentially growing yeast cells.

In brief, OxICAT uses isotope coded affinity tag (ICAT) technology [258-260] and is based on the rapid and irreversible modification of accessible cysteines in proteins with the thiol-trapping iodoacetamide (IAM) moiety of the ICAT reagent. In addition to the IAM moiety, the ICAT reagent consists of a cleavable biotin affinity tag and a 9-carbon linker which exists either in an isotopically light ^{12}C-form (herein referred to as light ICAT) or a 9 Da heavier isotopically heavy ^{13}C-form (herein referred to as heavy ICAT) [258]. In the first step of the OxICAT labeling process, proteins are denatured to gain access to all reduced cysteines, which are then irreversibly labeled with light ICAT (Fig. 3.2A). By using the strong thiol reductant Tris(2-carboxyethyl)phosphine (TCEP), reversible oxidative thiol modifications within the sample are then reduced and all newly accessible cysteines are subsequently modified with heavy ICAT (Fig. 3.2B). This sequential labeling with light and heavy ICAT generates chemically identical proteins, which only differ in the specific mass of their ICAT-label (9 Da additional mass per heavy ICAT), depending on their original redox state. After trypsin digest and affinity purification of the ICAT-labeled peptides, which markedly reduces the peptide complexity as we are only selecting for peptides that contain reactive cysteines, MS for quantification (Fig. 3.2, C and D) and tandem MS/MS for peptide identification is conducted.

OxICAT allows us to detect all reversibly oxidized cysteines, such as disulfide bonds, sulfenic acid formation or S-glutathionylation, whereas higher oxidation states, like sulfinic and sulfonic acids will not be detected. Peptides that contain originally only reduced cysteines are predicted to show single mass peaks, indicating the light ICAT-labeled form. Peptides containing originally only oxidized cysteines will generate masses that are exactly 9 Da, or in the case that more than one cysteine is oxidized in the peptide, multiples of 9 Da heavier than the corresponding light ICAT peptides. Most proteins with exposed redox-sensitive cysteine are in equilibrium with the prevailing redox conditions and are present in both oxidized and reduced form. Therefore, the corresponding peptides are present both in the light and heavy ICAT-modified form, forming an ICAT pair. Because light and heavy ICAT-labeled peptides are chemically identical, they will ionize to the same extent. This allows us to precisely determine their relative ion intensities, which correlate directly to their relative abundance. Therefore, the absolute ratio of oxidized and reduced peptide within a single sample can be calculated, which reflects the extent of oxidative thiol modifications present in the protein population. This makes OxICAT ideally suited not only for analyzing steady-state redox levels of proteins but also for monitoring changes in their redox status of cells exposed to stress conditions, because neither changes in protein expression nor protein stability influence the quantification of the OxICAT data.

MS/MS analysis of the peptide is then used to identify the respective protein and the oxidation-sensitive cysteine(s). To directly compare individual LC-MS runs, we used our in-house implemented OxICAT extension of *msInspect* [49]. This program automatically identifies and matches peptides across multiple LC runs and calculates precise changes in the oxidation status of hundreds of different proteins. *MsInspect* describes the series of parental mass spectra as image representation [251]. Whereas ICAT-labeled peptides are represented as black signals, ICAT-pairs are represented by two signals that show the same elution profile but are 9 Da (or multiples thereof) apart from each other on the *y*-axis (Fig. 3.2E).

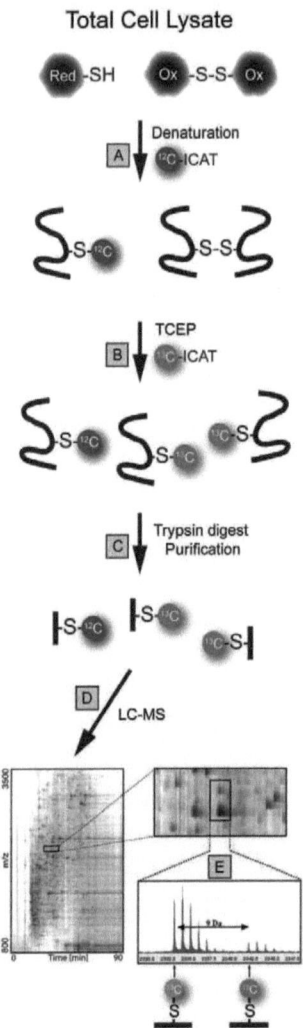

Figure 3.2. Scheme of the OxICAT method to quantify the oxidation state of protein thiols. A protein mixture of a total yeast cell lysate, in which proteins exist in either the reduced (left) or oxidized (right, here: disulfide-linked) form, is **(A)** incubated under denaturing conditions to expose all of its cysteine side chains. Then, isotopically light ^{12}C-ICAT reagent (green spheres) is added, which irreversibly modifies all reduced cysteines in the proteins. **(B)** All reversible oxidized thiol modifications are reduced with TCEP and subsequently all newly accessible cysteines are modified with isotopically heavy ^{13}C-ICAT reagent (red spheres). **(C)** The protein mixture is trypsin digested and ICAT-labeled peptides are purified by using the biotin-affinity tag. **(D)** Quantitative MS of the protein mixture reveals the extent of cysteine oxidation in any given peptide. Shown is the graphical representation of a LC-MS analysis of a total cell lysate of S. cerevisiae cells. The intensity of each mass signal is given as fraction of blackness and each signal corresponds to one either light or heavy OxICAT labeled peptide. **(E)** A peptide of a hypothetical protein elutes as ICAT pair with a mass difference of 9 Da. The mass signal with the lower m/z at 2333.24 Da has incorporated light ICAT, representing the reduced form of the peptide. The mass signal with the higher m/z at 2342.23 Da has incorporated heavy ICAT and represents the oxidized form of the peptide. Peptide sequence and identity of the modified cysteine are then determined by MS/MS.

3.3.2 Analysis of the *in vivo* redox status of organelle-specific proteins in yeast

A major problem for analyzing the redox state of organelle-specific proteins is the destruction of subcellular compartments by acid quenching, which is the necessary first step of any thiol trapping technique [261]. Addition of high concentrations of TCA (*i.e.*, 10%) to the cells is the standard procedure to accurately trap the *in vivo* thiol status of proteins and prevent unwanted oxidation and rearrangement processes. This treatment, however, inevitably destroys all subcellular compartments and results in the dilution of organelle-specific proteins with other cellular proteins. The dilution is further amplified by the circumstance that many organelle-specific proteins are of low abundance compared not only to many cytoplasmic proteins but also to a few very high abundant organelle-specific proteins. This decrease in concentration of organelle-specific proteins within total cell lysates during sample preparation makes their MS-based protein identification often impossible. Therefore, organelle purification is usually conducted to increase the protein yield and the chance for identification [255,256]. The problem arises, however, that the time-consuming organelle purification causes obvious changes in the thiol-oxidation status of organelle peptides [54,236]. Therefore, purified sub-cellular peptides, although always used, are less than optimal for the precise determination of the oxidation state. We now bypassed these obstacles that hampered the precise determination of the redox state of compartmental proteins by first labeling and identifying only purified organelle-specific proteins with light and heavy ICAT. This step allowed us to increase the MS/MS identification yield of organelle-specific peptides. We then used their specific *m/z*-values and positions in the LC run to identify sub-cellular peptides in our acid-trapped total cell lysates, which allowed us to precisely determine their redox status *in vivo*.

To obtain light and heavy labeled organelle-specific peptides, we purified the organelles of choice (*i.e.*, mitochondria, vacuoles, nuclei), and then applied a modified version of our thiol trapping technique to this organelle preparation. We reduced all cysteines in the denatured proteins with TCEP (Fig. 3.3A), split the samples and labeled one set with light ICAT to represent the reduced peptides, and the other one with heavy ICAT to represent the peptides whose cysteines were originally all oxidized (Fig. 3.3B). After trypsin digest and peptide purification, we mixed the labeled sets in a Heavy to Light ratio of 1:1 and subjected them to LC-MS/MS. With this method, we significantly increased the yield of identification of organelle-specific peptides (see below), and determined the *m/z* value for each peptide in the reduced (light ICAT) and oxidized (heavy ICAT) form and their precise LC-elution volume (Fig. 3.3C). We then used our OxICAT extension of *msInspect* [49] to match these peptides to the peptides of OxICAT-labeled total cell lysates (Fig. 3.4). Because LC-elution times can vary across multiple LC runs, which potentially results in a failure of peptide matching, we initially also spiked total cell lysates with heavy and light labeled organelle-specific peptides (Fig. 3.3D). This provided our software with the necessary information to assign the organelle-specific peptides then also to their counterparts in the non-spiked total cell lysates. Thus, we are now in the unique situation to precisely determine the *in vivo* oxidation state of organelle-specific proteins within their regular cellular environment (Fig. 3.4).

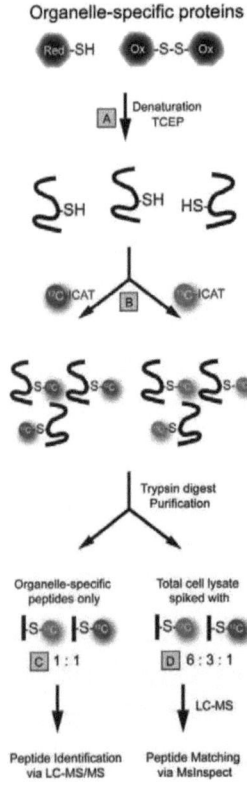

Figure 3.3. OxICAT principle for labeling of organelle-specific peptides. After organelle purification, the organelle-specific proteins, which exist in either the reduced (left) or oxidized form (right, here: disulfide-linked) are first **(A)** incubated under denaturing conditions to expose all of its cysteine side chains and reduced with TCEP. Then, the protein sample is split and the reduced cysteines are modified **(B)** either with isotopically light ^{12}C-ICAT reagent (green spheres) or heavy ^{13}C-ICAT reagent (red spheres). The light and heavy-labeled protein mixtures are trypsin digested and the ICAT-labeled peptides are purified by using the biotin-affinity tag. **(C)** For preparation of LC-MS/MS samples containing organelle-specific peptides, eluate of light and heavy labeled peptides were mixed in a 1:1 ratio, subjected to biotin tag cleavage and LC-MS/MS analysis. **(D)** For identification purposes in total cell lysates, lysates were supplemented with light and heavy labeled peptides in the ratio: total cell lysate/heavy labeled peptides/light labeled peptides of 6:3:1, subjected to biotin tag cleavage and analyzed using LC-MS.

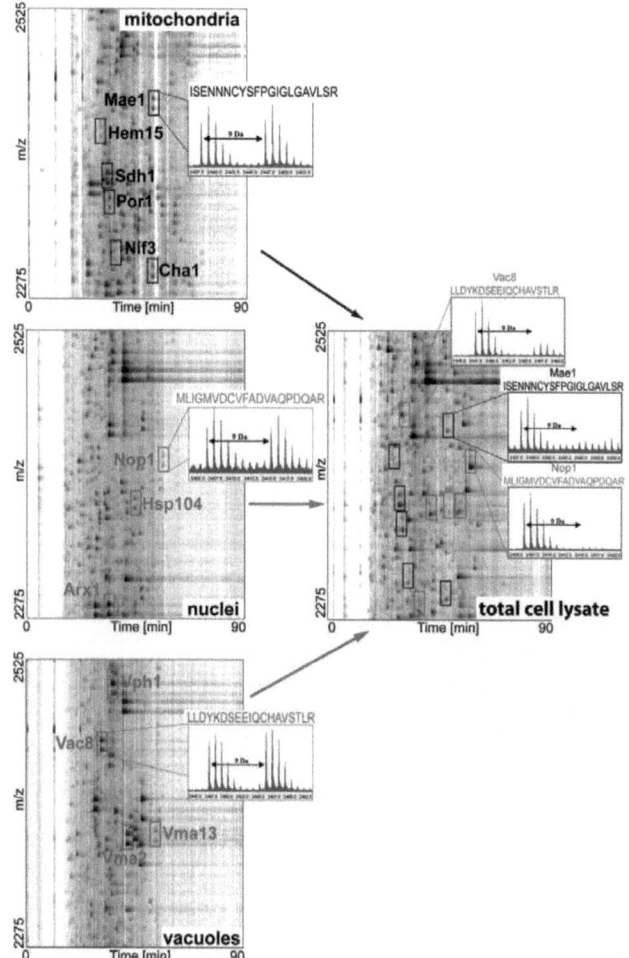

Figure 3.4. Using organelle-specific protein enrichment for enhanced peptide identification in total cell lysates. *After organelle purification, organelle-specific proteins are OxICAT labeled. Then, light and heavy labeled peptides are mixed in a 1:1 ratio (see Figures 3.2 and 3.3 for details) and LC-MS/MS is conducted on each individual organelle sample. On the left side, the graphical representations of the same 2275 Da to 2525 Da range of three different, organelle-specific LC-MS analyses (i.e., mitochondria, nuclei, vacuoles) are shown. In each analysis, up to six peptides per organelle are indicated by colored squares, which were identified via MS/MS analysis. For representative purposes, one peptide per organelle was selected and the respective mass spectrum is shown. The two mass peaks correspond to the light and heavy ICAT labeled peptides mixed in a 1:1 ratio. The right panel shows the same 2275 Da to 2525 Da range of a total cell lysate. The selected organelle-specific peptides are highlighted in the total cell lysate, and the mass spectra indicate their in vivo redox status.*

3.3.3 The majority of proteins are in their reduced state during exponential growth

Our modified OxICAT technique allowed us to determine the steady-state oxidation status of 391 cysteine-containing peptides representing 290 different yeast proteins in exponentially growing yeast cells. (Appendix, Table 3.A). Because our MS/MS analysis reveals the identity of the individual proteins but not their localization, we were unable to distinguish between different isoenzymes of a protein that are localized to different cellular compartments or proteins that cycle between different cellular localizations. For our data analysis, we therefore grouped proteins into those predicted or demonstrated to be predominantly localized in the cytoplasm, mitochondria, ER and Golgi, vacuoles and cytoplasmic proteins known to cycle in and out of the nucleus (cytoplasm/nucleus). The remaining proteins, which are predicted to be present in more than one compartment (*i.e.*, cytosol and mitochondria), were all grouped together (Fig. 3.5).

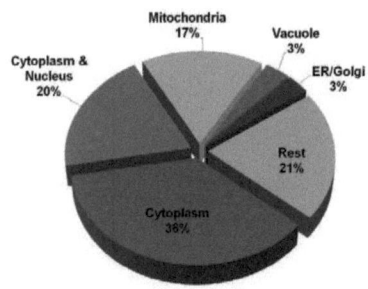

Figure 3.5. Cellular distribution of proteins in exponentially growing yeast cells identified by OxICAT. *Proteins were grouped according to their predominant cellular localization based on annotations in the S. cerevisiae Genome Database. Cytoplasmic proteins known to cycle in and out of the nucleus were grouped to Cytoplasm/Nucleus. Proteins with presence in more than one compartment (i.e., cytosol and mitochondria) were grouped together in "Rest".*

The activity of many cysteine-containing proteins, especially those localized to the cytoplasm, depends on preserving the thiol groups in the reduced state [51,262]. Thus, most cysteines in enzymes, whose activity is indispensable for cell viability, are predicted to be in their reduced form. Our data agreed with this assumption and showed that the majority of the identified peptides were almost exclusively labeled with light ICAT reagent (61% of total identified peptides of all compartments) and showed a level of oxidation of less than 15% (Fig. 3.6). These results are in excellent agreement with studies in *E. coli* that showed that the bulk of proteins are reduced [42,49,239,263] and are also in line with studies using a cytosol-based redox-sensitive YFP sensor, which was reported to show only a low extent of oxidation in exponentially growing yeast cells [239,264].

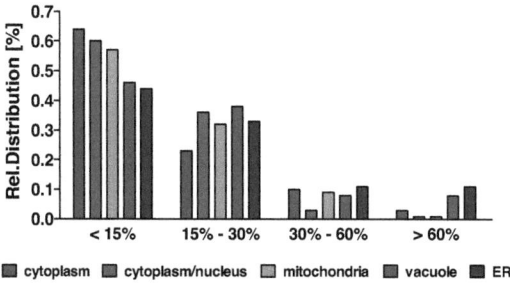

Figure 3.6. Distribution of organelle-specific peptides according to their oxidation state. *All ICAT-labeled peptides (see Appendix, Table 3.A) were grouped according to their cellular localization and oxidation status. Only peptides that are present in only one compartment were chosen for the analysis.*

The decline in the fraction of proteins below 15% oxidation appears to correlate with the increase in redox potential in the respective sub-cellular compartment. For instance, cytoplasm, and per extension the nucleus, exhibit the lowest, most reducing redox potentials (~ -290 mV) and the highest number of proteins with oxidation states below 15% oxidation. In contrast, vacuoles and the endoplasmic reticulum are considered to have the highest, most oxidizing redox potential in the cell (< -200 mV) and indeed are the two groups with the lowest number of largely reduced proteins (Fig. 3.6).

About 40% of all peptides harbored cysteines that were oxidized to more than 15% (Fig. 3.6). These results suggest that yeast cells are exposed to a considerable amount of ROS during logarithmic growth when glucose is used as primary carbon source. These results are in line with previous studies that demonstrated that mid-logarithmic phase yeast cells, cultivated aerobically in glucose, generate energy not exclusively by fermentation but also utilize oxidative phosphorylation, one of the main sources of cellular ROS [265-267]. It has been demonstrated that yeast cells shift their metabolism from fermentation to respiration when glucose in the media is completely depleted [265-267]. However, before glucose is completely exhausted, the degradation products ethanol and acetate are already used by oxidative phosphorylation in a metabolic phase called "respiro-fermentative growth" [265]. At the time of our thiol trapping experiment, we measured that about 22% of the initial glucose has been consumed (Fig. 3.7), suggesting that cells already transitioned to respiro-fermentative growth.

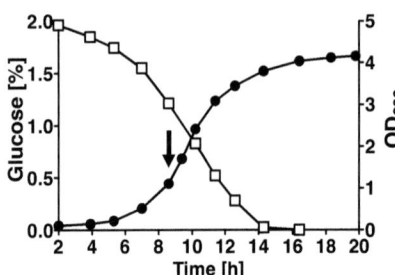

Figure 3.7. Optical density and glucose consumption of *S. cerevisiae* cultivated in 2% glucose medium. *Optical density at 600 nm (●) and glucose concentration in [%] (□) were measured in* S. cerevisiae *cultivated in 2% glucose medium. The arrow indicates the time when cells were harvested for OxICAT measurements.*

This suggests that sufficient ethanol and acetate is produced to enter the TCA cycle and initiate oxidative phosphorylation and ROS production. That the protein oxidation that we observe is likely caused by ROS derived from oxidative phosphorylation reactions became apparent when we compared the thiol redox status of proteins in yeast cells grown under largely fermentative conditions (i.e., 2% glucose) with cells cultivated on the non-fermentable carbon source of 3% glycerol. Under the latter conditions, cells generate energy only by respiration [268]. We observed that many proteins were significantly less oxidized in cells grown under largely fermentative conditions. For instance, Cys288 (aa 287-294) of the ADP, ATP carrier protein Pet9 was found to be oxidized to only 13% under fermentative conditions, while under respiratory conditions, 31% of Pet9 molecules have Cys288 oxidized. A similar trend was observed for cysteines of other proteins, including Cys124 (aa 123-130) of Kre2, Cys198 (aa 192-199) of Sah1, and Cys247 (aa 235-249) of Hom2. When cells are being grown with glucose as carbon source, cysteine oxidation was 33%, 16% and 13%, respectively. In contrast, the respective cysteine residues were oxidized to 47%, 24%, and 31%, respectively, when cells were cultivated in glycerol as non-fermentable carbon source.

These results were in excellent agreement with earlier studies in both yeast [245] and *E. coli* cells [42], which revealed significant thiol oxidation to several proteins during logarithmic growth. They suggested that ROS are constantly produced during aerobic growth and highlighted the essential roles that reducing and ROS-detoxifying systems, such as the thioredoxin and glutaredoxin systems play inside the cell [269].

3.3.4 Cu/Zn-Sod and Tim10 – Two highly oxidized proteins in yeast

Only about 5% of the peptides that we identified contained cysteines that were more than 60% oxidized under steady-state growth conditions (Fig. 3.6). Proteins with such oxidized cysteines included the Cu/Zn superoxide dismutase (Cu/Zn-Sod) and the mitochondrial import inner membrane translocase Tim10. Both proteins are known to form disulfide bonds as part of their catalytic function [270,271]. Superoxide dismutase catalyzes the disproportionation of superoxide anions to hydrogen peroxide and water [272] (Fig. 3.1). Yeast cells express two different superoxide dismutases, the mitochondrial Mn-Sod (product of the *sod2* gene) and Cu/Zn-Sod (product of the *sod1* gene, herein referred to as Sod1), which is mainly found in the cytosol but also in the mitochondrial intermembrane space [273]. In Sod1, an intramolecular disulfide bond between Cys58 and Cys147 has been reported as requirement for its activity in the cytoplasm [270,274]. Reduction of this disulfide converts the otherwise dimeric protein into a monomeric, inactive conformation [274]. We identified both of the cysteine-containing Sod1 peptides, aa 45-69 and aa 138-154, and found that the two cysteines, Cys58 and Cys147 are oxidized to 69% and 72%, respectively. These results strongly implied that the two cysteines are engaged in disulfide bond formation (Fig. 3.8) and that about 70% of Sod1 molecules are active under steady-state conditions *in vivo*. Our results are in agreement with previous studies that suggested that the majority of Cys58 and Cys147 in Sod1 is in its disulfide-bonded form in mid-log phase *in vivo* [270,275].

As reducing conditions prevail in the cytoplasm, the ~30% of reduced Sod1 likely reflects nascent Sod1 molecules [276]. In addition, since Sod1 is also found in the mitochondrial intermembrane space and mitochondrial import of the protein is significantly inhibited once the disulfide bond is introduced into the enzyme [277], the pool of reduced Sod1 molecules might represent a proportion of the protein that is taken up from the cytosol into the intermembrane space.

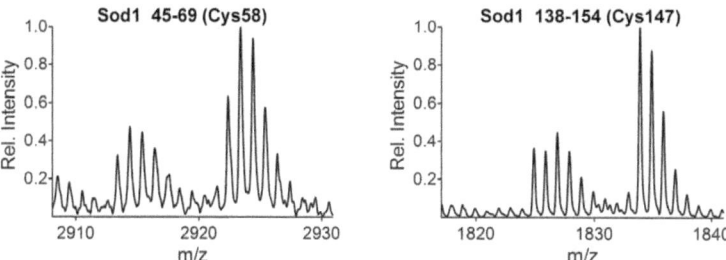

Figure 3.8. Cys58 and Cys147 in Sod1 are oxidized to the same extent. *Shown are details of the mass spectrum of the Sod1 peptides 45-69 and 138-154, which contain the redox-sensitive Cys58 and Cys147. The lower m/z peak represents the reduced, light-labeled form of the peptides*

(calculated m/z = 2913.35 and 1824.94, respectively), while the higher m/z peaks correspond to the oxidized, heavy-labeled forms of the peptides (calculated m/z = 2922.38 and 1833.97, respectively). The calculated extent of oxidation of Cys58 and Cys147 was 69% and 72%, respectively.

Tim10, which chaperones hydrophobic proteins inserted at the mitochondrial inner membrane, contains a twin CX_3C motif, with Cys44/Cys61 forming the inner disulfide bond and Cys40/Cys65 forming the outer disulfide bridge [271]. We identified that 93% of the peptide containing both Cys44 and Cys61 were labeled with two heavy ICAT labels, indicating that the two cysteines were almost completely oxidized *in vivo* (Fig. 3.9). This very high oxidation state confirms the crucial role of the disulfide bond for the function of Tim10, which has been shown to be essential for correct protein folding [271]. It also has a more prominent role in the folding process than the Cys40/Cys65 disulfide [271], which we did not identify in our OxICAT analysis.

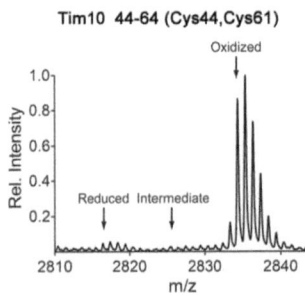

Figure 3.9. Cys44 and Cys61 in Tim10 are both highly oxidized. Shown are details of the mass spectrum of Tim10 peptide 44-64, which contains Cys44 and Cys64. The Tim10 peptide elutes as an ICAT-triplet; the fully reduced peptide, which is labeled with two light ICAT molecules (calculated m/z = 2816.28), the fully oxidized peptide, which is labeled with two heavy ICAT molecules (m/z = 2834.34) and an oxidation intermediate, in which one cysteine is labeled with the light ICAT and one cysteine is labeled with the heavy ICAT (calculated m/z = 2825.31). The two cysteines in peptide 44-64 are almost exclusively labeled with heavy ICAT, leading to a mass shift of 18 Da.

Apart from Sod1 and Tim10, we identified 12 other proteins with highly oxidized (> 60%) cysteines (Table 3.A Table 3.A). While some of these proteins have been previously shown to contain oxidation sensitive cysteines in yeast or other organisms (*e.g.*, Grx1, Ydj1), most have not been identified to form oxidative thiol modification under aerobic growth conditions *in vivo*. These proteins include several mitochondrial proteins (*e.g.*, Atp22, Trm1), vacuolar proteins (*e.g.*, Zps1, Ape3), ER-resident proteins (*e.g.*, Kre6, Fpr2) and several cytosolic proteins (*e.g.*, Fpr1, Gis2) including the ribosomal proteins Rpl40 and Rps29B. At this point, it is unclear what makes these cysteines particularly sensitive to oxidation, given that only very limited information is available for these proteins or homologues thereof. However, based on the fact that we did identify a number of proteins with known cysteine modifications, and confirmed the reduced nature of most other identified proteins, it is very likely that these proteins indeed either use oxidative thiol modifications as part of their catalytic function or for their structural stabilization. Noteworthy, several of the identified peptides contain two oxidized cysteines arranged in either a C-X-X-C motif (*e.g.,* Ydj1, Kre6, Grx1, Gis2), which is the typical cysteine arrangements in the active site of oxidoreductases and many metal-centers, or separated by more than two amino acids (*e.g.,* Zps1, Cys123 and Cys130; Rpl40A, Cys15 and Cys20; Pdh1, Cys65 and Cys72). This finding makes it very likely that these cysteines undergo disulfide bond formation *in vivo*. One cytosolic protein, for which we identified two peptides with two oxidized cysteines each, is the cellular nucleic acid binding protein Gis2, a homologue of human CNBP. Gis2 contains seven cysteine-rich CCHC putative zinc-finger motifs. The two peptides that we identified contain the C-X-X-C motifs of predicted zinc finger 3 (aa 48-59) and 6 (aa 118-129). The two cysteines in each peptide were oxidized to 66% and 70%, respectively. These results clearly suggest that at least two of the zinc finger motifs in Gis2 are sensitive to oxidative zinc release under aerobic growth conditions *in vivo*.

As the formation of disulfide bonds often causes structural rearrangements of the protein or works as regulatory instruments, detection of so far unknown disulfide bonds will aid in the understanding of the molecular mechanism of those proteins. At this point, we are unable to distinguish whether cysteine residues are highly oxidized because of their role in protein structure and/or function (*i.e.,* Sod1 or Tim10), or whether the cysteines are particularly sensitive towards ROS-mediated oxidation. The latter one seems less likely as glyceraldehyde-3-P dehydrogenase's (GapDH) active site cysteine Cys150, which is known to be highly oxidation-sensitive [51,245,257], was found to be only oxidized to about 26% under these aerobic growth conditions (Appendix, Table 3.A). Moreover, the extent of oxidative thiol modifications in many of these proteins was independent of the carbon source and energy metabolism (glucose vs. glycerol) suggesting that oxidation of these proteins occurs largely independent of ROS production. For instance, Cys44/Cys61 (aa 44-64) of Tim10 were oxidized to 93% / 83% (aerobic/glycerol), Cys147 (aa 138-154) of Sod1 to 72% / 66% and Cys672 (aa 664-677) of the mitochondrial translation factor Atp22 75% and 68%. These results suggest that these cysteines are highly oxidized because of their role in protein structure and/or function. If these cysteines were particularly sensitive toward ROS-mediated oxidation, an increase in their oxidation state under respiring conditions with glycerol as carbon source would have been expected. Under these conditions, increased oxidative phosphorylation results in increased levels of ROS compared to fermentative conditions with glucose as carbon source.

3.3.5 OxICAT to quantify the oxidation status of each cysteine separately in multi-cysteine proteins

Global methods that have been previously applied to detect oxidative thiol modifications in proteins are largely unable to distinguish between individual cysteine residues within the same protein. With OxICAT, we not only identify the affected cysteines but also quantify the extent of oxidation for each cysteine separately in multi-cysteine proteins. The individual quantification is especially interesting for cases where two or more cysteines within the same protein show different extents of oxidation. This should help us to find cysteines that function as catalytic or active site residues and play important roles for the activity of a protein. For instance, we found that in about 70% of Ydj1 molecules, 4 of the 8 conserved cysteines, which were predicted to constitute two zinc centers based on homology with bacterial DnaJ and human Hdj1, are largely oxidized. In contrast, however, the poorly conserved Cys370 in Ydj1 was found to be predominantly reduced. Another example is the ubiquitin-activating enzyme E1, for which we identified two cysteine-containing peptides. While Cys144 was found to be almost completely reduced (*i.e.,* 7% oxidation), Cys600 shows a moderate oxidation of 27%. Noteworthy, Cys600 represents the active site cysteine in E1, which forms a thio-ester bond with the C-terminus of ubiquitin [278], suggesting that OxICAT provides us a picture of the amount of enzymes that is in action. These examples show that the quantitative assessment of a protein's oxidation state is an extremely valuable tool to identify catalytically active and redox-sensitive cysteine.

3.3.6 Identification of oxidation-sensitive yeast targets

Cellular responses to hydrogen peroxide (H_2O_2) and superoxide ($\cdot O_2^-$, superoxide anion radical), have been extensively studied over the last decade and several redox-sensitive, cysteine-containing protein targets of oxidation-induced stress have been identified in *S. cerevisiae* and *E. coli* [49,262,279-281]. These proteins use the oxidation state of reactive cysteine residue as regulatory switches to modulate their activity. However, since these methods were mainly of qualitative nature, it is still unclear what proportion of these proteins undergo oxidative modifications under stress conditions and how physiologically significant this is. With our OxICAT technique, we have previously determined physiologically important target proteins of H_2O_2 and hypochlorite stress in *E. coli* and precisely quantified the extent of oxidative thiol modifications in these proteins [49]. Here, we focused on *S. cerevisiae* to identify physiologically relevant target proteins of defined ROS *in vivo* by determining the affected cysteine(s) and their precise oxidation status. These proteins have the potential to explain the physiological alterations that are observed in oxidatively stressed organisms. Moreover, they might function as marker proteins to detect type and onset of oxidative stress in aging organisms and numerous pathological conditions [281].

To identify peroxide and superoxide-sensitive proteins, we treated exponentially growing yeast cells in mid-logarithmic phase with concentrations of H_2O_2 and superoxide that transiently inhibited yeast growth (Fig. 3.10). Similar concentrations have been used in previous studies designed to investigate the response of yeast cells toward oxidative stress [262,279,280,282]. To detect proteins that are rapidly oxidatively modified *in vivo*, we harvested cells 15 min after addition of the respective oxidants and conducted our OxICAT thiol trapping experiments. We then compared the oxidation status of proteins before (see Chapter 4.3) and after oxidative stress treatment. Only those peptides, which showed at least a 1.4-fold increase in oxidative thiol modification upon exposure to oxidants in comparison to untreated samples in three or more independent experiments, were considered to be oxidation-sensitive.

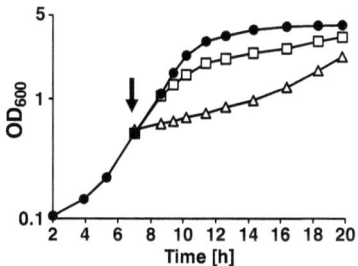

Figure 3.10. Growth of *S. cerevisiae* under oxidative stress conditions. S. cerevisiae EG103 *was grown in SDC until* OD_{600} *of 0.5 was reached. Then, the culture was split and either grown without additives (•) or in the presence of either 0.5 mM H_2O_2 (Δ) or 3 mM paraquat (□). The arrow indicates the addition of the oxidants.*

3.3.7 Translation, stress response and amino acid/carbohydrate metabolism are major targets of H_2O_2-mediated oxidation

While the majority of the 391 cysteine-containing peptides did not show any significant change in the extent of thiol modifications upon H_2O_2 exposure, 60 peptides, representing 55 different proteins displayed a significant increase in thiol oxidation (Appendix, Table 3.B). Among these proteins were a number of polypeptides which have been previously shown to become targets of H_2O_2-mediated thiol modifications, including GapDH, Ahp1, and pyruvate decarboxylase (see Appendix, Table 3.A for references) [51,245,283,284]. In contrast to these previous studies, which identified those redox-sensitive proteins but did not reveal their extent of oxidation or the cysteines involved, we have now a precise measurement of the degree of oxidation and the cysteine(s) modified. We found that the majority of identified proteins with H_2O_2–sensitive cysteines are involved in protein translation, amino acid and carbohydrate metabolism, and protein folding/stress response (Table 3.1). Analysis of the structure and function of these proteins suggested that cysteine modification should lead to a change in the function of most of these proteins. Noteworthy, H_2O_2 targets thiol groups of 14 proteins, whose genes are conditionally essential for growth of yeast cells [285]. Altering the functional state of one or more of these proteins by oxidative thiol modifications could be the cause for the observed growth inhibition of yeast cells upon treatment with H_2O_2.

Table 3.1. H$_2$O$_2$-sensitive proteins clustered in major biological processes

Process	Gene Name
Translation	DUS3, EFT2, ILS1, KRS1, RNA1, RPL12B, RPL21B, RPL37B, RPL40A, RPL42B, RPS11B, RPS31, RPS8B, SES1, TEF1, TEF4
Amino acid metabolism	ARO2, ARO4, ASN1, HOM2, HOM3, ILV6, SAH1, SHM1, THR1, TRP5
Carbohydrate and energy metabolism	ACO1, ADH1, FBA1, LPD1, PDC1, PGK1, TDH2
Protein folding and stress response	AHP1, CCT3, CCT7, CPR6, SSB2, YDJ1
Lipid metabolism	ERG10, ERG9, INO1
Protein degradation	CDC48, SHP1
Other metabolisms	ADH6, APA1, GUA1
Miscellaneous	CPR1, FRDS1, FUR1, MCM6, NOP1, SSO2, TFP1, YOR285W

3.3.8 Peroxide-mediated regulation of major functional classes

3.3.8.1 Carbohydrate metabolism

Many of the redox-regulated proteins that we identified with OxICAT turned out to be involved in carbohydrate metabolism (Table 3.1). A number of these proteins have previously been shown to harbor redox-sensitive cysteines, including GapDH or phosphoglycerate kinase 1 (Pgk1) [49,245]. GapDH undergoes significant increase in oxidative thiol modification at Cys150 and Cys154 upon H$_2$O$_2$ exposure (Appendix, Table 3.B). We were unable to distinguish which of the several GapDH isoenzymes are affected, as the identified peptide is identical in both Tdh2 and Tdh3 (Tdh1 is not expressed when cells are cultivated with glucose). During H$_2$O$_2$ exposure, both Tdh2 and Thd3 are known to be carbonylated [279], but only Tdh3 was found to be S-thiolated [286]. Previous studies suggested that redox regulation of GapDH mainly involves S-glutathionylation of its active site cysteine Cys150 [40,122]. However, based on our studies, the majority of oxidized GapDH molecules appear to be modified at both conserved cysteines 150 and 154, and only to a minor degree, at Cys150 alone (Fig. 3.11). This result strongly suggests that disulfide bond formation is the predominant redox regulatory mechanism of GapDH during oxidative stress. This makes excellent sense given that Cys154 is very close in structure and likely reacts with Cys150 to form a stable disulfide bond. Similar observations have been made with E. coli GapDH [40,122].

Disulfide bond formation of GapDH upon oxidative stress has been suggested to promote nuclear accumulation and neuronal cell death [48,112]. Besides GapDH, oxidation of a number of other redox-regulated proteins involved in carbohydrate metabolism has been previously linked to their inactivation [40]. Inactivation of these enzymes leads to a rerouting of the metabolic flux into the pentose phosphate pathway (PPP) and to increased NADPH production [108] (Fig. 3.12). Therefore, oxidative inactivation of glycolytic enzymes such as GapDH might function as a rapid strategy to increase NADPH generation under oxidative stress conditions and to restore redox homeostasis.

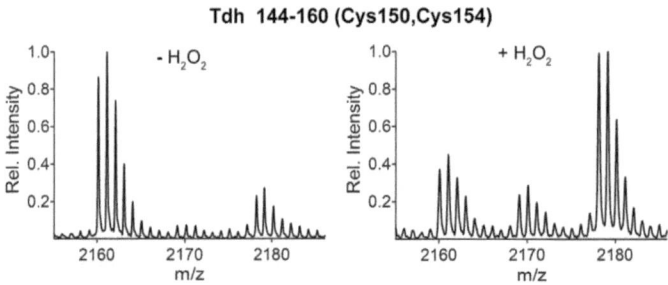

Figure 3.11. GapDH's Cys150 and Cys154 are both oxidation-sensitive. *Shown are details of the mass spectrum of Tdh2 or Tdh3 peptide 144-160, which contains the oxidation-sensitive Cys150 and Cys154, before (left side) and after addition of 0.5 mM H_2O_2 (right side) to the cell culture. The Tdh2/3 peptide elutes as an ICAT-triplet; the fully reduced peptide, which is labeled with two light ICAT molecules (m/z = 2160.11), the fully oxidized peptide, which is labeled with two heavy ICAT molecules (indicating oxidation at both cysteines; m/z = 2178.18) and an oxidation intermediate, in which one cysteine is labeled with the light ICAT and one cysteine is labeled with the heavy ICAT (m/z = 2169.15).*

In contrast to previous studies, we did not confirm the redox sensitivity of either one of the two cysteines in triosephosphate isomerase Tpi1 and we found only one (i.e., Cys112) of the five cysteines in fructose bisphosphate aldolase Fba1 to be moderately oxidation sensitive (Appendix, Table 3.A and 3.B) [51,245,279]. These results, which suggest that both Tpi1 and Fba1 remain active during H_2O_2 stress make physiological sense as increasing concentrations of DHAP and FBP, the respective products of Tpi1 and Fba1, further favor the production of NADPH by the pentose phosphate pathway (PPP) [287].

The low but significant levels of oxidation of active site cysteines that we observed in a large number of other proteins involved in glucose metabolism might likely play a physiological role in yeasts metabolism. Oxidation would lead to the reversible inactivation of some of the enzyme's population, which might contribute to the Pasteur effect that describes the decrease in glycolytic flux under aerobic growth conditions in yeast [288,289]. This nicely illustrates how reversible oxidative thiol modifications can be used as elegant mechanism to fine-tune cellular metabolism.

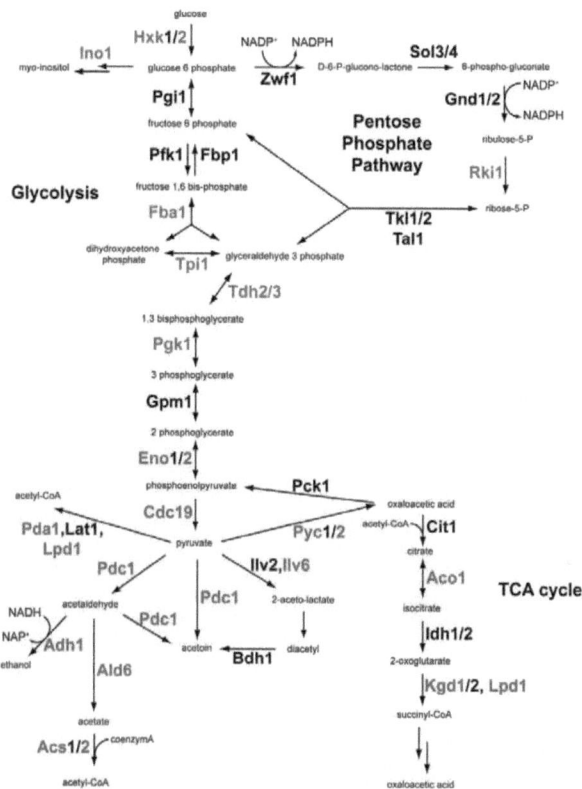

Figure 3.12. General overview of proteins containing H_2O_2-sensitive cysteines involved in carbohydrate & energy metabolism and pentose phosphate pathway. *Proteins in red were identified to be sensitive toward exogenous H_2O_2 exposure, those in green to be insensitive. Proteins in black were not identified. "/" denotes that two isoenzymes are involved in this step. Pathways are simplified due to space restriction.*

3.3.8.2 Amino acid biosynthesis

We identified several metabolic enzymes involved in amino acid biosynthesis that contained H_2O_2 sensitive cysteines, including tryptophan synthase Trp5, aspartokinase

Hom3, DAHP synthase Aro4, and asparagine synthetase Asn1. In *E. coli*, tryptophan synthase, which catalyzes the final two steps in tryptophan biosynthesis, has been shown to possess three cysteines (Cys62, Cys170, Cys230), which react with various sulfhydryl reagents [290]. Both Cys170, which is suggested to be redox-active in *S. typhimurium* [291] and Cys230, which corresponds to Cys529 in yeast [257] are important residues for the catalytic property of Trp5 in *S. typhimurium* [291,292]. We identified peptides containing the non-conserved residues Cys321 and Cys420 in yeast Trp5 to be redox-sensitive upon H_2O_2 stress in yeast, but were unable to identify any of the other cysteine-containing peptides from Trp5. The fact that we did observe some of Trp5's cysteines to be redox-sensitive suggests however that tryptophan synthesis is also redox-regulated in yeast. For Aro4, one of the first enzymes involved in the biosynthesis of aromatic amino in yeast [293], we showed that Cys76 is significantly sensitive towards H_2O_2 treatment. This residue corresponds to the active site residue in *Corynebacterium glutamicum*, whose mutation has been found to cause loss in activity [294].

3.3.8.3 Protein translation

Another group of proteins that appears to be targeted by H_2O_2 stress concerns proteins involved in protein translation. We identified several small and large ribosomal subunits, elongation factors and tRNA synthetases that exhibit a significant increase in oxidative thiol modifications after short-term exposure to H_2O_2 (Table 3.1). The elongation factor 1-alpha Ef1α, which binds to and delivers aminoacyl-tRNA to the A-site of ribosomes during protein biosynthesis and is encoded by the tef1 and tef2 genes, has previously been identified as H_2O_2-sensitive protein [245,257,263]. While we found Cys324, Cys368 and Cys409 in Ef1α to be H_2O_2-sensitive, McDonagh *et al.* [257] identified Cys409 in Ef1α as oxidation-sensitive residue. Modifications of proteins involved in protein translation are likely to down-regulate ribosomal activity during H_2O_2 treatment, which agrees well with our observations (Sebastian Ahrends, personal communication) and that of others that protein translation is reduced during peroxide stress [295,296]. Inhibition of ribosomal activity will save energy resources needed to fight oxidative damage and prevent the production of nascent polypeptides under potentially error-prone conditions, which might rapidly fall victim to potentially irreversible oxidative damage.

3.3.8.4 Stress Response and Protein Folding

Elevated levels of ROS require cells to take immediate action to destroy the oxidants to prevent oxidative damage. One of these ROS scavengers is the thiol-specific antioxidant alkyl hydroperoxide reductase 1 (Ahp1), which uses the catalytic Cys62 for H_2O_2 and alkyl hydroperoxide detoxification and the resolving Cys120 for cysteine regeneration [297,298]. We observed only a very low level of thiol oxidation of Ahp1's active site cysteine in unstressed cells (Fig. 3.13, A and B, left panel) suggesting that Ahp1 is mainly active upon exogenous H_2O_2 treatment, and not during exponential growth when peroxide levels are significantly lower.

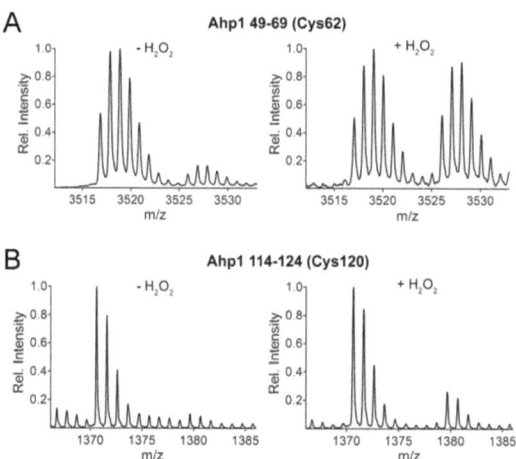

Figure 3.13. Cys62 and Cys120 in Ahp1 are both oxidation-sensitive. *Shown are details of the mass spectrum of the Ahp1 peptides (A) 49-69 and (B) 114-124, which contain the redox-sensitive (A) Cys62 and (B) Cys120, before (left side) and after addition of 0.5 mM H_2O_2 (right side) to the cell culture. The lower m/z peak represents the reduced form of the peptides, while the higher m/z peak corresponds to the respective oxidized forms of the peptides.*

We found that after exogenous H_2O_2 treatment, oxidation of cysteine Cys62 and Cys120 increases about 3-fold (Fig. 3.13, A and B, right panel; Appendix, Table 4.B). This is in excellent agreement with the formation of an intermolecular disulfide bond between Cys62 and Cys120 [298]. Our data appear to furthermore reflect the kinetics of peroxide detoxification, as Cys62, the primary site of oxidation by H_2O_2, exhibits a higher degree of oxidation than Cys120, which serves as the revolving cysteine (Fig. 3.13, compare A and B). Cys31, the third cysteine in Ahp1 that is not part of peroxide detoxification, does not show any change in oxidative thiol modification upon H_2O_2 exposure (Appendix, Table 4.A). These findings reiterate the power of the OxICAT method not only for detecting redox-sensitive proteins but also to locate the redox-sensitive cysteines within those proteins.

Stress conditions often lead to the unfolding of proteins, which then require the help of chaperones for their refolding into their native conformation [157]. In our study, we identified several H_2O_2-sensitive proteins that are involved in protein folding (*i.e.*, Ssb2, Cpr6, Ydj1) (Table 3.1). It has been shown in *A. thaliana* that the conserved Cys20 of the heat-shock protein 70 chaperone Ssb2 is mandatory for its chaperone function under oxidative stress, although the exact role of Cys20 remains to be demonstrated [299]. Vignols *et al.* speculate that Cys20 may be engaged in a disulfide bond with a cysteine of another Hsp70 molecule [299]. Although we did not find the Cys20-containing peptide in our study, we did identify the only other cysteine in Ssb2, Cys454, to be redox-sensitive. This cysteine might represent the disulfide partner of Cys20 in Ssb2.

It remains to be investigated how functional changes of Ssb2 during oxidative stress affect the chaperone's function. Oxidative thiol modifications in proteins that are sensitive towards H_2O_2 could change their affinity to unfolded proteins or cause the rearrangement of their conformation. It is reasonable to assume that such modifications would alter their chaperone activity, thus affecting cell survival.

3.3.9 Localization of H_2O_2-sensitive target proteins

Our studies revealed that proteins with peroxide-sensitive cysteines were found in almost all cellular compartments, but seemed to be overrepresented in the cytoplasm and underrepresented in the mitochondria. (Fig. 3.14). Mirzaei *et al.* [300] showed that oxidative amino acid modifications (*e.g.*, aldehyde and ketone formation) upon H_2O_2 exposure were randomly distributed inside the cell. Such random distribution can be expected given that we generated widespread exogenous peroxide stress. The bias toward cytoplasmic proteins might be explained with the fact that cytoplasmic proteins are probably the first proteins to encounter H_2O_2 upon entering the cell. Mitochondria, however, have additional membranes and intrinsic enzymes that fight ROS. Oxidation caused by endogenous ROS might be more localized toward the mitochondria, as mitochondria are considered to be the major source of ROS inside the cell.

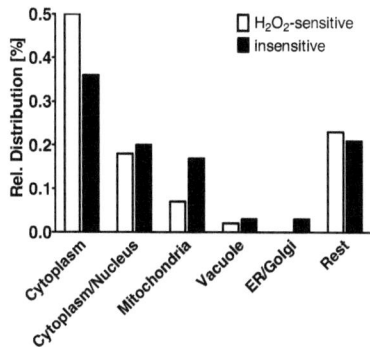

Figure 3.14. Cellular distribution of H_2O_2-insensitive proteins compared to H_2O_2-sensitive proteins. *Proteins were grouped according to their predominant cellular localization based on annotations in the S. cerevisiae Genome Database (SGD). Cytoplasmic proteins known to cycle in and out of the nucleus were grouped to Cytoplasm/Nucleus. Proteins with presence in more than one compartment (i.e., cytosol and mitochondria) were grouped together in "Rest".*

3.3.10 Structural analysis reveals no differences between H_2O_2-sensitive and insensitive cysteine residues

To obtain insights into structural or sequence parameters that might make particular cysteines more oxidative-stress sensitive, we compared pK_a values and structural accessibility of the identified H_2O_2-sensitive and insensitive cysteines. Also, we predicted the secondary structure and fold propensity of the amino acid stretches these cysteine residues are located in. The reactivity of a cysteine is influenced primarily by its pK_a value, which is influenced by its structural environment [30]. The known crystal structures of 20 proteins with peroxide-sensitive cysteines and 72 proteins with H_2O_2-insensitive cysteines were used for our calculations. Cysteines that showed more than 35% oxidation during logarithmic phase were excluded to make sure only thiols whose oxidation was caused by exogenous addition of H_2O_2 were statistically analyzed. Using the program PROPKA, we calculated the pK_a values of the respective cysteines and found that the average pK_a of H_2O_2-sensitive cysteine residues was significantly lower than the pK_a value of H_2O_2-insensitive residues (t-test, P = 0.043) (Fig. 3.15A). This result was in excellent agreement with previous observations that pK_a values of redox-sensitive cysteines are lower compared to non-reactive cysteine residue [31,32]. It has been shown that the majority of cytoplasmic protein thiols have pK_a values of greater than 8.0, which render the thiol groups predominantly protonated and largely non-reactive at intracellular pH [30]. Regulatory thiol modifications of redox-sensitive cysteines, however, have characteristically lower pK_a values [31,32]. Under physiological pH conditions, these thiols are therefore present as deprotonated, highly reactive thiolate anions (RS⁻). The low pK_a values of redox-sensitive cysteines are thought to arise primarily from stabilizing charge-

charge interactions between the thiolate anion and neighboring positively charged or aromatic side chains [33,34]. These thiolate anions, in contrast to their protonated counterparts, are highly susceptible to ROS-mediated oxidation [35]. Noteworthy, we found several H_2O_2 sensitive cysteines with pK_a values larger than 8 (Fig. 3.15A).

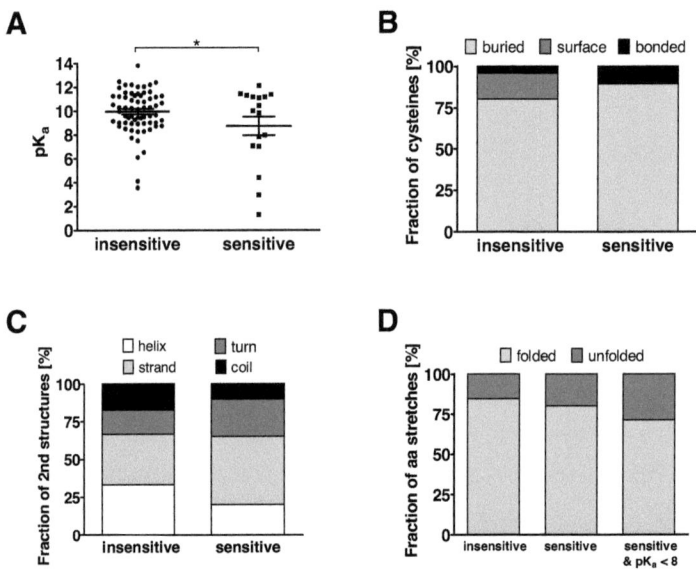

Figure 3.15. Structural analysis between H_2O_2-sensitive and insensitive cysteine residues. Cysteine residues were analyzed for **(A)** pK_a value, **(B)** cysteine accessibility, **(C)** secondary structure of the cysteine-containing amino acid stretch and **(D)** fold propensity of the amino acid stretch that the cysteine is located in.

We assume that these cysteines represent the non-reactive cysteine partner of the disulfide bonds. In proteins like the members of the thioredoxin family, which use reversible disulfide bond formation as part of their catalytic cycle, it has been shown that one of the two cysteines is reactive with an unusually low pK_a value while the other cysteine is non-reactive and is suggested to exert a pK_a values not below 11 [301]. We also detected several redox-insensitive cysteines with pK_a values of less than 8. It is conceivable that these cysteines are buried and therefore inaccessible under physiological conditions.

As previously mentioned, one often cited theory states that the high reactivity and unusual pK_a value derives from stabilizing interactions of the thiolate anion with neighboring positively charged or aromatic amino acids [33,34]. Preliminary sequence and structural comparison of redox-sensitive and insensitive cysteines did not confirm this theory suggesting that other features might be involved as well. We therefore investigated accessibility, secondary structure features and fold predictions to assess what makes certain cysteines so highly reactive. Again, we used the known crystal structures of the 20 H_2O_2-sensitive and 72 insensitive cysteines that were identified with OxICAT for these predictions. Using the program PROPKA, we were unable to detect any significant differences in the accessibility of cysteine between the group of H_2O_2-sensitive and the group of insensitive cysteines (Fig. 3.15B). We did observe, however, that in the group of peroxide-insensitive cysteines, more cysteine residues (chi-test, $P = 0.087$) were found to be on the surface of the protein. These results suggest that a higher accessibility of the cysteine residue does not translate into a redox-sensitive cysteine. Since we are analyzing oxidative modifications that occur within and are stable over a 15 minute incubation time, these results do not exclude that surface accessible cysteines are oxidized but rapidly reduced by cellular oxidoreductases.

Predictions of the type of secondary structure the H_2O_2-sensitive or insensitive cysteines are located in by STRIDE resulted in no significant difference between the two groups (Fig. 3.15C). For both groups, the majority of cysteines were found to reside in α-helices and β-strands, and a smaller number in turns and coils. This might indicate that the flexibility of the secondary structure has only little to no effect on the redox-sensitivity of cysteines. Also, no significant difference was observed for the prediction of the fold propensity of the amino acid stretch that contains either a redox-sensitive or insensitive cysteine. FoldIndex predicted a similar number of folded and unfolded stretches in the group of H_2O_2-sensitive cysteines and in the group of insensitive cysteines (Fig. 3.15D). When we performed the prediction of the fold propensity only with those seven H_2O_2-sensitive cysteines that show lower than normal pK_a values (n = 7, average pK_a of 5.40), we found that about 30% of these cysteines reside in unfolded regions as compared to 16% of H_2O_2-insensitive cysteines (Fig. 3.15D). Despite the low n-number, these results suggest that the fold propensity might be one of the few parameters that could determine the redox sensitivity of cysteines.

3.3.11 Aconitase as major target of superoxide stress in yeast

While endogenous superoxide is mainly produced in the mitochondria, we used exogenous treatment with paraquat to generate intracellular superoxide. Superoxide production by paraquat occurs in the mitochondrial matrix, where paraquat undergoes redox cycling by being reduced by an electron donor such NADPH, before being oxidized by an electron receptor such as oxygen to produce superoxide [302]. Upon sublethal paraquat treatment of yeast cells for 15 min, we identified 8 peptides that show an increase in oxidative thiol modifications compared to untreated cells. This result indicates that paraquat-mediated thiol modification of proteins affects significantly fewer proteins than H_2O_2 does. This result is in excellent agreement with previous reports that show that the number of oxidative thiol modifications is smaller in paraquat/superoxide-treated cells as compared to H_2O_2 [280,303]. For instance, H_2O_2 treatment in *B. subtilis* allowed for the identification of 25 thiol-containing proteins, whereas paraquat treatment affected only six protein [303]. Interestingly, neither one of these six proteins contained an 4Fe-4S cluster although proteins with 4Fe-4S cluster have been suggested to be the primary targets of superoxide treatment [304,305]. In our study, we detected aconitase 1 (Aco1) as the only 4Fe-4S cluster containing protein sensitive towards superoxide treatment. We identified Aco1's Cys382, which is part of the 4Fe-4S cluster, as the major target site of superoxide-mediated oxidation. These results suggest that the oxidation of other, non-Fe-S cluster containing superoxide-sensitive proteins may either reflect an indirect effect, for instance a decrease of cytoplasmic reductants such as NADPH when electrons are transferred to paraquat, or might be caused by the formation of ROS like H_2O_2, which are generated during the detoxification process of superoxide. Alternatively, oxidation of 4Fe–4S clusters, such as in aconitase, causes the release of iron [306]. Free iron promotes the formation of highly reactive hydroxyl radicals via Fenton chemistry, which can cause extensive oxidative damage in the cell and contribute to mitochondrial dysfunction (Fig. 3.1) [307,308]. Progressive mitochondrial damage leads to the complete inability to respire and ultimately cell death, as respiration is essential for survival in stationary phase.

Five of the eight superoxide-sensitive proteins are involved in amino acid metabolism. One of the main reasons for cells to inactivate these proteins could be the prevention of wasteful energy consumption by less important cellular processes such as amino acid biosynthesis in times of elevated stress levels. Oxidation of aconitase, an important component of the TCA cycle, might result in a reduction of the metabolic flux through the

TCA. Thus, cells would possess a quick mechanism to redirect energy and metabolites toward increasing NADPH production in the PPP. Noteworthy, we found that paraquat-induced superoxide generation also causes oxidation of 6-phosphogluconate dehydrogenase (Gnd1), which catalyzes the second NADPH regenerating reaction in the pentose phosphate pathway (Fig. 3.12) [309]. In contrast, glucose-6-phosphate dehydrogenase (Zwf1), which catalyzes the first irreversible and rate-limiting step of the pentose phosphate pathway, also regenerates NADPH [310] and is not oxidized. Additionally, Gnd2, which encodes for the minor isoform of phosphogluconate dehydrogenase and is responsible for about 20% of Gdn's activity [309], is also not affected by H_2O_2 and might take over some additional activity from a potentially partly inactivated Gnd1.

Interestingly, only three proteins harbor cysteine residues that are sensitive towards both H_2O_2 and superoxide treatment (Appendix, Table 4.B). Additionally, one protein, tryptophan synthase Trp5, has two cysteines of which one exhibits sensitivity towards H_2O_2-mediated oxidation while the other one is sensitive to superoxide treatment. The partial overlap in oxidation sensitivity might be either due to a 1) dual sensitivity of the respective cysteines to these two, seemingly unrelated oxidants, 2) of the conversion of superoxide to hydrogen peroxide by superoxide dismutases (Fig. 3.1) [272]. If the latter was the case, however, more than only three proteins would be expected to be oxidized by both H_2O_2 and superoxide unless the conversion from superoxide to H_2O_2 was very localized or happened very gradually and affected only the most peroxide sensitive cysteines.

5 out of 8 superoxide-sensitive peptides differ from the H_2O_2-sensitive cysteine targets. These results indicate that the two physiological oxidants have a distinct set of *in vivo* target proteins. This specificity might be explained with the distinct reactivity of thiol groups in individual proteins towards these oxidants and could reflect the specific mode of action of the two oxidants. Superoxide production by paraquat occurs in the mitochondrial matrix, where paraquat has to undergo redox cycling for superoxide generation [302,311]. Superoxide also carries a negative charge and cannot easily pass through membranes, while H_2O_2 is a non-charged molecule, which rapidly diffuses into cells. Noteworthy, the differences in the cellular distribution of H_2O_2- and superoxide-affected proteins that we observed in our studies might reflect the chemical difference of these oxidants. H_2O_2-

sensitive proteins were found in almost all compartments with a small bias toward the yeast cytoplasm, whereas 5 of the 8 superoxide-sensitive proteins were localized in the mitochondria. This result is in excellent agreement with the fact that mitochondria are the central superoxide-generating organelles not only during the paraquat treatment but also during respiratory growth.

Interestingly, exposure of cells to paraquat does not immediately result in growth inhibition (Fig. 3.10) [312]. In our OxICAT experiments, paraquat was added during logarithmic phase where about 75% of the initial glucose was still present in the media and cells grew mainly fermentative. As fermentative cell growth was not inhibited, these results suggest that superoxide did not target any essential proteins involved in fermentative growth. This result agreed with our finding that none of the previously identified H_2O_2-sensitive and essential proteins were affected by superoxide treatment. A visible inhibition of cell growth became apparent about 2 hours after the paraquat treatment, coinciding with the depletion of glucose in the media and the switch to respiratory growth. Importantly, this is the growth phase where cells require a functional aconitase, one of the main targets of superoxide treatment. Inactivation of aconitase results in the decrease in respiration, and likely explains the defect in cell growth.

The recovery of yeast cells from superoxide treatment, which is observed several hours after the treatment, could be explained with the repair of oxidized Fe-S-cluster containing proteins, such as aconitase. It has been shown that a superoxide-inactivated form of dihydroxyacid dehydratase and aconitase in *E. coli* can be restored in the presence of crude *E. coli* extract, iron and sulfide after the removal of the oxidant [312,313]. This indicates that the proteins are still intact and can be reactivated by rebuilding the cluster.

3.4 Conclusions

In the past decade, it has been shown that reversible oxidative thiol modifications such as disulfide bonds, sulfenic acid formation or glutathionylation regulate the activity of many oxidation-sensitive pro- and eukaryotic proteins [51,225,314]. These proteins participate in various cellular processes including translation, glycolysis, stress protection, signal transduction, and amino acid biosynthesis, making these pathways redox-controlled [38,70,225]. Numerous studies have also shown a clear link between protein oxidation and aging, neurodegenerative diseases or diabetes [5,315-318]. In the past, most proteomic approaches have solely aimed at identifying proteins with oxidized thiols while the quantitative aspect of cysteine oxidation was mostly neglected for technical reasons. This piece of information is, however, crucial to assess the effects of oxidative thiol modifications on metabolic and signal transduction pathways [50,319,320].

We have now established a global thiol trapping technique termed OxICAT to quantify the redox state of hundreds of different proteins. We have determined the *in vivo* oxidation status of proteins residing in different cellular compartments in exponentially growing cells and identified oxidation-sensitive cysteines in many of these proteins. Based on the reported structural or functional role of cysteines in many of these proteins, we can now predict how changes in their redox state will affect the structure and function of these proteins.

Most of the proteins that we identified to harbor oxidized cysteines can be clustered in three groups: 1) proteins that are reduced in unstressed cells but have redox-sensitive cysteine(s) that affect the catalytic function of the protein (*e.g.*, aconitase) during oxidative stress; 2) proteins with highly oxidation-sensitive active site cysteines (*e.g.*, GapDH), which are partially oxidized in exponentially growing yeast cells due to their exquisite sensitivity towards low amounts of ROS; 3) proteins with structurally or functionally important disulfide bonds (*e.g.*, Tim10, Sod1).

We found that the majority of identified proteins in this study display a low oxidation state. This was expected during logarithmic growth in glucose where most of the energy is derived from fermentative processes. Noteworthy, some of the oxidation-sensitive proteins were identified to exhibit elevated degrees of oxidation-mediated thiol modifications. At least 30% of these proteins turned out to be proteins harboring H_2O_2 sensitive cysteines, suggesting that yeast cells might already encounter a certain degree of ROS during logarithmic growth under fermentative conditions. These results highlight the importance of cellular thiol-disulfide oxidoreductases, such as the glutaredoxin and thioredoxin system, which maintain the cellular redox balance at the expense of NADPH.

We identified several proteins that either have been previously shown (Sod1, Trr1/2, and Ahp1) or were predicted (Ydj1, Zps1) to undergo oxidation-mediated thiol modification *in vivo*. This makes us very confident that thiol modifications of other cysteine-containing proteins discovered in this study also play regulatory or functional roles. To our knowledge, this analysis provides us now with the most extensive and quantitative list of redox-sensitive proteins in yeast cells. The next step will be to conduct detailed biochemical analysis to investigate the exact role that thiol modifications play in some of these potentially redox-regulated proteins. In the past, it has been shown that many transcriptional (e.g., OxyR, Yap1) [70,321] and posttranslational (e.g., Hsp33, Prx-2) [225,298] responses are triggered by reversible thiol modifications.

Many of the oxidation-sensitive proteins identified in our analysis were found to be involved in protein translation and ribosome biogenesis. This is in excellent agreement with several recent studies, which showed inhibition of protein synthesis and the increased oxidation of translation factors in aged or oxidation-compromised bacterial and yeast cells [37,51,322]. This result suggests that proteins important for translation efficiency and fidelity might be one of the major targets of oxidative stress. Other major targets of peroxide stress in the cell are enzymes of glycolysis, TCA cycle and amino acid biosynthesis. Together, this data indicate that cells redirect the metabolic flux and energy away from glycolysis, translation and biosynthesis of amino acids toward the pentose phosphate pathway. This would allow the constant production of the reducing equivalent NADPH, which is used by enzymes of the antioxidant defense as important electron donor. These results show how cells employ posttranslational thiol modifications for a coordinated metabolic response to assure their survival by counteracting the effects of peroxide stress.

To assess what makes cysteines so particularly sensitive to oxidation, we investigated whether we detect any sequence or structural features that are common among these H_2O_2-sensitive cysteines and less represented in cysteines that are oxidation resistant. We analyzed pK_a values and structural accessibility of the cysteines, and predicted secondary structure and fold propensity of the amino acid stretches in which the cysteines are located in. The only parameter that we identified to be significantly different in oxidation-sensitive cysteines as compared to resistant cysteines is the fold propensity. Cysteines that are located in unfolded regions appear to be more prone to oxidation than cysteines that are located in a folded region. Interestingly, the redox-sensitive cysteines of other known redox-regulated proteins such as the chaperone Hsp33 or the transcription factors RsrA, OxyR and Yap1 are all located in unfolded regions. Oxidation of the redox-sensitive cysteine residues usually results in the activation of these proteins [29,43,50], which is often accompanied by structural rearrangements of the affected protein [29,50]. It is feasible to argue that these rearrangements are less likely to take place in folded regions, where interactions between amino acid residues might prevent such events.

It can also be speculated that cysteines in unfolded regions are more likely to be exposed to ROS than cysteines in folded regions, where the folded structure might function as protective measure against ROS. It has been shown that the protein thiol pool is an active player in the antioxidant defense, as protein thiols consumed more than half of the oxidizing equivalents upon diamide exposure [323]. Also, electron-rich side chains of many amino acid residues such as Met, His, Trp, and Tyr are suggested to react with oxidizing reactive species [324,325]. Thus, the oxidative load on individual amino acids might be higher in unfolded regions than in folded stretches, where neighboring amino acids could participate in ROS scavenging. However, such ROS scavenging in folded regions could be disadvantageous for enzymes that are involved in physiological signaling processes, where low ROS concentrations are used for signal transduction [5,6,12,13]. Thus, it could be hypothesized that redox-sensitive cysteines of proteins which are involved in antioxidant defense or signal transduction might preferentially be located in unfolded protein regions. However, no ROS-specific/mediated activity has been shown yet for the proteins identified in this study that contain H_2O_2-sensitive cysteine residues in unfolded regions.

A feature that truly distinguishes OxICAT from other redox proteomic approaches is the ability to precisely quantify the extent of oxidative thiol modifications for each individual cysteine identified by mass spectrometry. However, with our chosen approach that uses TCEP as reductant, we were not able to identify the type of reversible thiol modification. TCEP works as an unspecific reductant that provides us only with a general overview of all reversible thiol modifications *in vivo*. To detect specific modifications, specialized reductants can be used, such as ascorbic acid to reduce nitrosothiols [197], sodium arsenite for sulfenic acids [284], or glutaredoxins to reduce glutathionylated cysteines [326]. Also, even though we were able to increase the number of proteins identified by organelle purification, we are still unable to identify low abundance yeast proteins, a common theme in all proteomics approaches conducted so far. Thus, future methods should aim to increase the sensitivity of redox proteomics to low abundance proteins, such as the known redox-regulated proteins Yap1 or the redox transducer Gpx3 [70,73].

Oxidation-mediated thiol modifications are attributed to many physiological and pathological conditions such as cancer and neurodegenerative diseases [22,23], and are thought to be the underlying culprit of eukaryotic aging. Knowledge of oxidation-sensitive proteins and their thiol status might help to understand these diseases and improve their therapies, such as the development of more effective or new antioxidants. Additionally, oxidation-sensitive proteins could be used as marker proteins to detect type and onset of oxidation during cellular aging. It has been shown that several targets of chronological aging overlap with proteins that exhibit sensitivity toward different oxidants [281]. Also, the identification of the same or related oxidized proteins in studies across different species [37,49,327] indicates remarkable specificity and conservation of protein thiol oxidation and highlights thiol oxidation as important regulatory mechanism throughout evolution.

3.5 Appendix Chapter 3

Table 3.A. Thiol oxidation status of yeast proteins during exponential growth. OxICAT was applied to S. cerevisiae whole cells and purified mitochondria, vacuoles, and nuclei. Identified peptides are grouped by their gene name. Gene name, protein name, and Swiss-Prot accession numbers are based on annotations in SGD (www.yeastgenome.org). "% oxidation" denotes the average oxidation of the redox-sensitive cysteine (second column, in brackets), obtained from at least three independent experiments ± standard deviation. The localization C, cytoplasm; M, mitochondria; N, nucleus; G, golgi; V, vacuoles; ER, endoplasmic reticulum is according to the component listed in SGD. §: essential gene; *: peptide identical in both isoenzymes; a,b,c: at this mass, two different peptides were identified.

Swiss-Prot	Gene, Peptide (Cys)	Protein name	Cellular compartment	% oxidation during exponential growth	Reference
Q08641	ABP140, 431-461 (442)	Uncharacterized methyltransferase	C	16 ± 5	
Q00955	ACC1, 1212-1241 (1225)	Acetyl-CoA carboxylase	ER	16 ± 3	
P19414	ACO1, 117-134 (122)	Aconitase 1	C, M	15 ± 2	[42,245,262,284,327]
P19414	ACO1, 376-392 (382)	Aconitase 1	C, M	7 ± 2	[42,245,262,284,327]
P39533	ACO2, 136-157 (154)	Aconitase 2	M	21 ± 3	
P39533	ACO2, 60-81 (66)	Aconitase 2	M	23 ± 8	
P52910	ACS2, 197-206 (201)	Acetyl-coenzyme A synthetase 2	C, N	13 ± 3	
P60010	ACT1, 216-238 (217)	Actin	C, N	24 ± 4	[263,281,326]
P38009	ADE17, 285-304 (300)	Bifunctional purine biosynthesis protein	C	8 ± 4	
P38009	ADE17, 361-378 (362)	Bifunctional purine biosynthesis protein	C, N	18 ± 5	
P07245	ADE3, 603-621 (612)	C-1-tetrahydrofolate synthase	C	13 ± 6	
P38972	ADE6, 591-609 (607)	Phosphoribosylformylglycinamidine synthase	C	11 ± 5	
P00330	ADH1, 277-299 (277,278)	Alcohol dehydrogenase 1	C	17 ± 5	[42,51,245,263,281,328]
Q04894	ADH6, 141-174 (163)	NADP-dependent alcohol dehydrogenase 6	C	24 ± 3	
P47143	ADO1, 121-130 (121)	Adenosine kinase	C, N	11 ± 4	[245]
P38013	AHP1, 114-124 (120)	Peroxiredoxin type-2	C	6 ± 2	[245,298,328]
P38013	AHP1, 16-32 (31)	Peroxiredoxin type-2	C	13 ± 3	[245,298,328]
P38013	AHP1, 49-79 (62)	Peroxiredoxin type-2	C	14 ± 2	[245,298,328]
P36112	AIM28, 529-538 (535)	Uncharacterized	M	20 ± 4	
P46367	ALD4, 61-80 (74)	Aldehyde dehydrogenase (K+ dep.)	M	18 ± 2	
P54115	ALD6, 124-134 (132)	Aldehyde dehydrogenase (Mg2+ dep.)	C, M	15 ± 4	[42,51,245,246]
P54115	ALD6, 290-311 (306)	Aldehyde dehydrogenase (Mg2+ dep.)	C, M	15 ± 4	[42,51,245,246]
P54783	ALO1, 359-374 (359)	D-arabinono-1,4-lactone oxidase	M	22 ± 4	
P16550	APA1, 127-139 (129)	AP4A phosphorylase	C, N	7 ± 5	
P16550	APA1, 292-321 (306)	AP4A phosphorylase	C, N	13 ± 2	
P37302	APE3, 179-199 (187)	Aminopeptidase Y	V	76 ± 7	[257]
P46672	ARC1, 261-274 (266)	GU4 nucleic-binding protein 1	C	17 ± 9	

Chapter 3: Oxidation-sensitive S. cerevisiae proteins

UniProt	Gene, position	Protein name	Loc.	Value	Ref.
Q05933	ARC18, 32-52 (47)	Actin-related protein 2/3 complex subunit 3	M	56 ± 7	
P27768	ARG1, 283-294 (284)	Argininosuccinate synthase	C	6 ± 2	
P28777	ARG2, 210-224 (221)	Chorismate synthase	C	13 ± 3	
P32449	ARO4, 240-249 (244)	Phospho-2-dehydro-3-deoxyheptonate aldolase	C, N	13 ± 6	
P32449	ARO4, 69-90 (76)	Phospho-2-dehydro-3-deoxyheptonate aldolase	C, N	9 ± 2	[257]
Q03862	ARX1, 58-75 (65)	Probable metalloprotease ARX1	C, N	6 ± 5	
P49089	ASN1, 2-15 (2)	Asparagine synthetase 1	C	17 ± 3	
P49089	ASN1, 421-440 (432)	Asparagine synthetase 1	C	10 ± 4	
P49089/P49090	ASN1/2, 396-406 (404)*	Asparagine synthetase 1/2	C	7 ± 2	
P50273	ATP22, 664-677 (672)	Mitochondrial translation factor ATP22	M	75 ± 6	
P38077	ATP3, 115-126 (117)	ATP synthase subunit gamma	M	9 ± 4	
P38891	BAT1, 125-141 (126)	Branched-chain-amino-acid aminotransferase	M	11 ± 5	[37]
P38891	BAT1, 220-237 (227)	Branched-chain-amino-acid aminotransferase	M	8 ± 4	
P29311	BMH1, 77-88 (85)	14-3-3 protein	C, N	16 ± 4	
P53890	BNI5, 253-269 (266)	Bud neck protein 5	C	25 ± 4	
P39077	CCT3, 512-531 (518)§	T-complex protein 1 subunit gamma	C	14 ± 3	
P39078	CCT4, 393-402 (399)	T-complex protein 1 subunit delta	C	12 ± 6	[37]
P39079	CCT6, 515-535 (530)	T-complex protein 1 subunit zeta	C	5 ± 1	
P39079	CCT6, 79-103 (96)	T-complex protein 1 subunit zeta	C	10 ± 4	
P42943	CCT7, 452-469 (454) §	T-complex protein 1 subunit eta	C	11 ± 3	
P47079	CCT8, 335-343 (336)	T-complex protein 1 subunit theta	C	47 ± 10	
O13547	CCW14, 38-61 (42;51;53)	Covalently-linked cell wall protein 14	C, M	93 ± 4	
P00549	CDC19, 287-309 (296)§	Pyruvate kinase 1	C	12 ± 5	[42,245,326,328]
P00549	CDC19, 370-394 (371)§	Pyruvate kinase 1	C	18 ± 5	[42,245,326,328]
P00549	CDC19, 414-425 (418)§	Pyruvate kinase 1	C	46 ± 7	[42,245,326,328]
P25694	CDC48, 104-119 (115)§	Cell division control protein 48	ER, N	12 ± 3	[245]
P25694	CDC48, 76-93 (92)§	Cell division control protein 48	ER, N	15 ± 8	[245]
P25379	CHA1, 175-195 (178)	L-serine/threonine dehydratase	M	11 ± 3	
P25379	CHA1, 257-271 (269)	L-serine/threonine dehydratase	M	3 ± 1	
P25379	CHA1, 59-80 (78)	L-serine/threonine dehydratase	M	23 ± 3	
Q01519	COX12, 42-49 (48)	Cytochrome c oxidase subunit 6B	C	12 ± 3	
P07258	CPA1, 8-27 (11)	Carbamoyl-phosphate synthase small chain	C	24 ± 7	[257]
P03965	CPA2, 159-182 (170)	Carbamoyl-phosphate synthase large chain	C	10 ± 5	
P03965	CPA2, 673-694 (678)	Carbamoyl-phosphate synthase large chain	C	18 ± 2	
P14832	CPR1, 36-53 (39)ª	Peptidyl-prolyl cis-trans isomerase CPR1	C	12 ± 3	[245,246,279]
P53691	CPR6, 150-165 (156)	Peptidyl-prolyl cis-trans isomerase CPR6	C	7 ± 2	[245,246]
P53691	CPR6, 270-282 (279)	Peptidyl-prolyl cis-trans isomerase CPR6	C	3 ± 1	[245,246]
P27614	CPS1, 361-370 (368)	Carboxypeptidase S	V	33 ± 3	
P38845	CRP1, 105-118 (113)	Uncharacterized	N	21 ± 3	
P00175	CYB2, 508-524 (521)	Cytochrome b2	M	15 ± 2	
Q32582	CYS4, 300-322 (301)	Cystathionine beta-synthase	C, M	10 ± 4	[245]
P24783	DBP2, 215-226 (219)	ATP-dependent RNA helicase DBP2	C, N, M	15 ± 2	
P20449	DBP5, 156-172 (167)	ATP-dependent RNA helicase DBP5	C, N	17 ± 3	
P06634	DED1, 256-270 (257)	ATP-dependent RNA helicase DED1	C	11 ± 5	

ID	Gene, residues	Protein	Loc.	Value	Ref.
P06634	DED1, 275-285 (276)	ATP-dependent RNA helicase DED1	C	14 ± 3	
P46681	DLD2, 198-214 (200)	D-lactate dehydrogenase 2	M	21 ± 4	
P39976	DLD3, 408-426 (409)	D-lactate dehydrogenase 3	C	13 ± 5	
P14020	DPM1, 168-175 (172)	Dolichol-phosphate mannosyltransferase	C, M, ER	14 ± 4	
P04802	DPS1, 250-259 (255)	Aspartyl-tRNA synthetase	C, N	28 ± 4	
Q06053	DUS3, 210-225 (224)	tRNA-dihydrouridine synthase 3	C, N	13 ± 2	
P32324	EFT2, 353-370 (366)[b]	Elongation factor 2	R	12 ± 4	
P32324	EFT2, 121-144 (136)	Elongation factor 2	C	11 ± 2	
P32324	EFT2, 441-465 (448)	Elongation factor 2	C	8 ± 5	
P32324	EFT2, 724-749 (735)	Elongation factor 2	R	8 ± 4	
P00925	ENO2, 244-255 (248)	Enolase 2	C, V, M	5 ± 1	[51,263,281,283,328]
P98333	ENP1, 320-333 (330)	Essential nuclear protein 1	C, N	11 ± 3	
P32476	ERG1, 170-189 (174)	Squalene monooxygenase	ER	27 ± 4	
P41338	ERG10, 343-362 (358)[a,§]	Acetyl-CoA acetyltransferase	C	12 ± 3	
P10614	ERG11, 137-146 (142)	Cytochrome P450	ER	15 ± 6	
P54839	ERG13, 289-301 (300)	Hydroxymethylglutaryl-CoA synthase	ER	17 ± 6	
P25087	ERG6, 120-135 (128)	Sterol 24-C-methyltransferase	ER, M	18 ± 4	
P29704	ERG9, 340-350 (341)[§]	Squalene synthetase	ER, M	9 ± 2	
P47912	FAA4, 26-52 (49)	Long chain fatty acid CoA ligase 4	C	4 ± 3	
P47912	FAA4, 400-414 (403)	Long chain fatty acid CoA ligase 4	C	9 ± 4	
P07149	FAS1, 1296-1317 (1308)	Fatty acid synthase subunit beta	C, M	13 ± 4	
P19097	FAS2, 471-481 (480)	Fatty acid synthase subunit alpha	C, M	9 ± 4	[262]
P19097	FAS2, 900-919 (917)	Fatty acid synthase subunit alpha	C, M	22 ± 3	[262]
P14540	FBA1, 133-160 (158)[§]	Fructose-bisphosphate aldolase	C, M	26 ± 2	[51,245,263,281,326,328]
P14540	FBA1, 209-230 (226)[§]	Fructose-bisphosphate aldolase	C, M	12 ± 3	[51,245,263,281,326,328]
P14540	FBA1, 250-264 (254)[§]	Fructose-bisphosphate aldolase	C, M	13 ± 3	[51,245,263,281,326,328]
P14540	FBA1, 286-308 (292)[§]	Fructose-bisphosphate aldolase	C, M	8 ± 4	[51,245,263,281,326,328]
P14540	FBA1, 96-115 (112)[§]	Fructose-bisphosphate aldolase	C, M	13 ± 2	[51,245,263,281,326,328]
P38631	FKS1, 1047-1063 (1056)	1,3-beta-glucan synthase component	M	31 ± 5	
P20081	FPR1, 50-64 (55)	FK506-binding protein 1	C, N, M	76 ± 6	
P32472	FPR2, 30-39 (36)	FK506-binding protein 2	ER	79 ± 3	
P32614	FRDS1, 446-459 (454)	Probable fumarate reductase	C, M	6 ± 2	
P15624	FRS1, 121-138 (137)	Phenylalanyl-tRNA synthetase beta chain	C	12 ± 5	
P18562	FUR1, 163-172 (170)	Uracil phosphoribosyltransferase	C	10 ± 4	
P18562	FUR1, 195-216 (215)	Uracil phosphoribosyltransferase	C	11 ± 6	
Q05670	FUS2, 366-375 (371)	Nuclear fusion protein FUS2	unknown	18 ± 4	
P09032	GCD1, 163-188 (172)	Translation initiation factor eIF-2B subunit gamma	N	13 ± 3	
P33892	GCN1, 1070-1078 (1077)	Translational activator GCN1	C, M	22 ± 6	
P39958	GDI1, 217-226 (221)	Rab GDP-dissociation inhibitor	C	14 ± 4	
Q12154	GET3, 314-322 (317)	ATPase GET3	ER	15 ± 4	
P25370	GFD2, 88-103 (90)	Good for DBP5 activity protein 2	unknown	16 ± 7	
P53849	GIS2, 118-129 (118;121)	Zinc finger protein GIS2	C	70 ± 5	
P53849	GIS2, 48-59 (49;52)	Zinc finger protein GIS2	C	66 ± 7	

ID	Gene, residues	Protein	Value	Loc.	Refs
P53849	GIS2, 67-78 (67;70)	Zinc finger protein GIS2	52 ± 6	C	
P32598	GLC7, 20-40 (38)	Serine/threonine-protein phosphatase	9 ± 4	N	
P17709	GLK1, 440-460 (448)	Glucokinase 1	15 ± 6	C	
P32288	GLN1, 149-170 (160)	Glutamine synthetase	15 ± 4	C	
P38720	GND1, 287-295 (287)	6-phosphogluconate dehydrogenase	15 ± 4	C, M	
P41277/P40106	GPP1/GPP2, 210-222 (211;211)*	(DL)-glycerol-3-phosphatase 1/2	27 ± 7	C, N	[245]
Q12068	GRE2, 77-105 (86)	NADPH-dependent methylglyoxal reductase	5 ± 2	C, N	
P25373	GRX1, 25-45 (27;30)	Glutaredoxin 1	80 ± 10	C, N	
P38625	GUA1, 438-455 (450)	GMP synthase	9 ± 3	unknown	
P46655	GUS1, 333-344 (342)	Glutamyl-tRNA synthetase	6 ± 3	C, M	
P46655	GUS1, 378-400 (385)	Glutamyl-tRNA synthetase	15 ± 5	C, M	
P09950	HEM1, 139-151 (145)	5-aminolevulinate synthase	19 ± 4	M	
P09950	HEM1, 386-391 (386)	5-aminolevulinate synthase	13 ± 4	M	
P16622	HEM15, 118-136 (123)	Ferrochelatase	27 ± 4	M	
P00498	HIS1, 149-179 (160)	ATP phosphoribosyltransferase	9 ± 2	C	
P00498	HIS1, 71-82 (71)	ATP phosphoribosyltransferase	17 ± 2	C	
P00815	HIS4, 449-472 (470)	Histidine biosynthesis trifunctional protein	6 ± 2	C	
P40037	HMF1, 106-129 (107)	Protein HMF1	16 ± 4	C, N	
P13663	HOM2, 235-249 (247)	Aspartate-semialdehyde dehydrogenase	13 ± 3	M	[42,245]
P13663	HOM2, 250-263 (259)	Aspartate-semialdehyde dehydrogenase	21 ± 3	C, N	[42,245]
P10869	HOM3, 38-56 (53)	Aspartokinase	5 ± 1	C	
P31539	HSP104, 387-407 (400)	Heat shock protein 104	11 ± 4	C, N	
P19882	HSP60, 127-138 (132)	Heat shock protein 60	20 ± 3	M	[245,246,262,281,326]
P19882	HSP60, 266-287 (283)	Heat shock protein 60	18 ± 3	M	[245,246,262,281,326]
P33416	HSP78, 325-337 (333)	Heat shock protein 78	13 ± 3	M	
P04807	HXK2, 254-282 (268)	Hexokinase 2	21 ± 3	C, N, M	[245,298]
P04807	HXK2, 395-407 (398,404)	Hexokinase 2	17 ± 3	C, N, M	[245,298]
P21954	IDP1, 360-371 (370)	Isocitrate dehydrogenase 1	10 ± 5	M	[42,327]
P21954	IDP1, 393-403 (398)	Isocitrate dehydrogenase 1	18 ± 4	M	
P21954	IDP1, 89-98 (89)	Isocitrate dehydrogenase 1	13 ± 3	M	[42,327]
P09436	ILS1, 99-120 (119)§	Isoleucyl-tRNA synthetase	25 ± 5	C	
P00927	ILV1, 126-146 (132)	Threonine dehydratase	14 ± 3	M	
P00927	ILV1, 304-332 (321)	Threonine dehydratase	14 ± 4	M	
P39522	ILV3, 157-180 (176)	Dihydroxy-acid dehydratase	14 ± 4	M	
P06168	ILV5, 307-321 (316)	Ketol-acid reductoisomerase	15 ± 3	M	[245]
P25605	ILV6, 102-120 (111)	Acetolactate synthase small subunit	8 ± 2	M	
P25605	ILV6, 236-254 (244)	Acetolactate synthase small subunit	15 ± 3	M	
P25605	ILV6, 261-272 (270)	Acetolactate synthase small subunit	9 ± 4	M	
P25605	ILV6, 76-93 (82)	Acetolactate synthase small subunit	7 ± 3	M	
P50095	IMD3, 247-263 (250)	Inosine-5'-monophosphate dehydrogenase	14 ± 3	C	[326]
P11986	INO1, 19-29 (19)	Inositol-1-phosphate synthase	34 ± 4	C	
P40069	KAP123, 308-339 (310)	Importin subunit beta-4	3 ± 3	C, N	
P40069	KAP123, 977-987 (984)	Importin subunit beta-4	14 ± 3	C, N	
P22147	KEM1, 1494-1512 (1508)	5'-3' exoribonuclease 1	16 ± 3	C	

Chapter 3: Oxidation-sensitive *S. cerevisiae* proteins 109

Accession	Gene, positions	Description	Loc.	Value	Ref.
P20967	KGD1, 812-834 (816)	2-oxoglutarate dehydrogenase E1	C, M	22 ± 4	
P20967	KGD1, 982-996 (983)	2-oxoglutarate dehydrogenase E1	M	25 ± 3	
P27809	KRE2, 123-130 (124)	Glycolipid 2-alpha-mannosyltransferase	G	33 ± 3	
P32486	KRE6, 476-493 (479;481)	Beta-glucan synthesis-associated protein	G, ER	92 ± 3	
P15180	KRS1, 429-446 (438)§	Lysyl-tRNA synthetase	C	9 ± 4	
P15180	KRS1, 206-229 (212)§	Lysyl-tRNA synthetase	C	12 ± 6	
P15180	KRS1, 430-446 (438)§	Lysyl-tRNA synthetase	C	10 ± 5	
P15180	KRS1, 481-496 (486)§	Lysyl-tRNA synthetase	C	13 ± 2	
P14904	LAP4, 191-203 (202)	Vacuolar aminopeptidase 1	V	19 ± 3	
P07264	LEU1, 230-242 (233)	3-isopropylmalate dehydratase	C	15 ± 4	[245]
P07264	LEU1, 60-77 (63)	3-isopropylmalate dehydratase	C, N	22 ± 4	[245]
P06208	LEU4, 345-353 (345)	2-isopropylmalate synthase	C, M	10 ± 3	
P09624	LPD1, 45-58 (54)	Dihydrolipoyl dehydrogenase	M	12 ± 2	
P53312	LSC2, 365-377 (365)	Succinyl-CoA ligase subunit beta	M	12 ± 4	
P07702	LYS2, 611-625 (614)	L-aminoadipate-semialdehyde dehydrogenase	C	15 ± 7	
Q12121	LYS21, 194-206 (202)	Homocitrate synthase	N	13 ± 4	
Q12122	LYS21, 289-319 (314)	Homocitrate synthase	N	14 ± 6	
P49367	LYS4, 335-344 (340)	Homoaconitase	M	35 ± 5	
P38999	LYS9, 333-342 (340)	Saccharopine dehydrogenase	C	15 ± 4	[328]
P38999	LYS9, 4-34 (33)	Saccharopine dehydrogenase	C	11 ± 3	[328]
P36013	MAE1, 202-221 (220)	Malic enzyme	M	29 ± 5	
P36013	MAE1, 535-555 (541)	Malic enzyme	M	10 ± 3	
P36013	MAE1, 88-111 (96)	Malic enzyme	M	19 ± 3	
P29469	MCM2, 677-684 (681)	DNA replication licensing factor MCM2	C, N	14 ± 3	
P53091	MCM6, 813-832 (815)§	DNA replication licensing factor MCM6	C, N	20 ± 5	
P36060	MCR1, 256-277 (263)	NADH-cytochrome b5 reductase 2	M	34 ± 5	
Q12283	MCT1, 197-211 (198)	Malonyl CoA-acyl carrier protein transacylase	M	27 ± 5	
P00958	MES1, 353-366 (353)	Methionyl-tRNA synthetase	C	23 ± 5	
P32179	MET22, 339-355 (349)	3'(2'),5'-bisphosphate nucleotidase	C	20 ± 2	
P05694	MET6, 563-573 (571)	Methionine synthase	C	14 ± 3	[42,245]
P05694	MET6, 650-667 (657)	Methionine synthase	C	12 ± 7	[42,245]
P09440	MIS1, 216-229 (224)	C-1-tetrahydrofolate synthase	M	10 ± 3	
P53152	MMS2, 81-101 (85)	Ubiquitin-conjugating enzyme MMS2	C, N	17 ± 5	
P53875	MRPL19, 36-46 (40)°	54S ribosomal protein L19	M	7 ± 3	
P36517	MRPL4, 62-76 (62)	54S ribosomal protein L4	M	9 ± 3	
P25293	NAP1, 231-250 (249)	Nucleosome assembly protein	C, M	16 ± 6	
P53081	NIF3, 105-123 (114)	NGG1-interacting factor 3	C, M	10 ± 3	
P15646	NOP1, 232-251 (239)§	rRNA 2'-O-methyltransferase fibrillarin	C, N	7 ± 5	
P15646	NOP1, 273-288 (275)§	rRNA 2'-O-methyltransferase fibrillarin	C, N	9 ± 3	
P15646	NOP1, 312-319 (314)§	rRNA 2'-O-methyltransferase fibrillarin	C, N	9 ± 7	
P27476	NSR1, 233-254 (241)	Nuclear localization sequence-binding protein	N, M	14 ± 6	
Q02630	NUP116, 1030-1042 (1031)	Nucleoporin NUP116	N	9 ± 6	
Q03028	ODC1, 215-240 (230)	Mitochondrial 2-oxodicarboxylate carrier 1	M	29 ± 6	

Chapter 3: Oxidation-sensitive *S. cerevisiae* proteins

UniProt	Peptide	Protein	Loc.	Value	Ref.
P38219	OLA1, 126-143 (126)	Uncharacterized GTP-binding protein	C	5 ± 3	
P38219	OLA1, 305-319 (313)	Uncharacterized GTP-binding protein	C	15 ± 6	
P38219	OLA1, 43-62 (43)	Uncharacterized GTP-binding protein	C	17 ± 7	
P38219	OLA1, 72-81 (76)	Uncharacterized GTP-binding protein	C	9 ± 3	
P38325	OM14, 28-35 (29)	Mitochondrial outer membrane protein	M	4 ± 2	
Q12447	PAA1, 48-61 (51,55)	Polyamine N-acetyltransferase 1	C	26 ± 8	
P04147	PAB1, 363-377 (368)	Polyadenylate-binding protein	C, N	18 ± 4	
P16387	PDA1, 140-147 (140)	Pyruvate dehydrogenase E1 subunit alpha	M	15 ± 3	
P06169	PDC1, 128-154 (152)	Pyruvate decarboxylase isozyme 1	C, N	16 ± 3	[245,262,263,281]
P06169	PDC1, 213-224 (221;222)	Pyruvate decarboxylase isozyme 1	C, N	22 ± 4	[245,262,263,281]
P16467	PDC5, 128-154 (152)	Pyruvate decarboxylase isozyme 2	C, N	11 ± 8	
Q12428	PDH1, 64-78 (65;72)	Probable 2-methylcitrate dehydratase	C, M	11 ± 2	
P33302	PDR5, 1048-1054 (1050)	Pleiotropic ABC efflux transporter	M	30 ± 6	
P18239	PET9, 66-76 (73)ᶜ	ADP,ATP carrier protein 2	M	7 ± 3	
P18239	PET9, 265-273 (271)	ADP,ATP carrier protein 2	M	16 ± 4	[327]
P18239	PET9, 287-294 (288)	ADP,ATP carrier protein 2	M	13 ± 2	[327]
P18239	PET9, 67-76 (73)	ADP,ATP carrier protein 2	M	13 ± 2	[327]
P07274	PFY1, 77-91 (89)	Profilin	C	7 ± 4	
P00560	PGK1, 83-108 (98)ˢ	Phosphoglycerate kinase	C, M	9 ± 3	[37,245,246,263,328,329]
P37012	PGM2, 372-389 (376)ᵇ	Phosphoglucomutase-2	C	12 ± 6	
P34217	PIN4, 338-355 (343)	RNA-binding protein PIN4	C	12 ± 6	
P05030/P19657	PMA1/2, 392-414 (376-405)*	Plasma membrane ATPase 1/2	C, M	13 ± 3	
P05030/P19657	PMA1/2, 415-443 (409-438)*	Plasma membrane ATPase 1/2	C, M	12 ± 7	
P05030/P19657	PMA1/2, 487-503 (472;501)*	Plasma membrane ATPase 1/2	C, M	18 ± 5	
P04840	POR1, 125-132 (130)	Mitochondrial outer membrane protein porin 1	M	8 ± 4	[245,328]
P04840	POR1, 206-224 (210)	Mitochondrial outer membrane protein porin 1	M	8 ± 3	[245,328]
P53043	PPT1, 244-262 (247)	Serine/threonine-protein phosphatase T	C, N	12 ± 8	
P09232	PRB1, 344-368 (361)	Cerevisin	V	28 ± 5	
P21242	PRE10, 73-86 (76)	Proteasome component C1	N	12 ± 4	
P41940	PSA1, 100-125 (113)	Mannose-1-phosphate guanyltransferase	C	12 ± 3	[245,263]
P32379	PUP2, 94-122 (117)	Proteasome component PUP2	C, N	23 ± 10	
P07275	PUT2, 154-171 (162)	Delta-1-pyrroline-5-carboxylate dehydrogenase	M	12 ± 4	
P32327	PYC2, 211-222 (218)	Pyruvate carboxylase 2	C	11 ± 2	
P07256	QCR1, 302-323 (312)	Cytochrome b-c1 complex subunit 1	M	26 ± 3	
P07257	QCR2, 129-152 (137)	Cytochrome b-c1 complex subunit 2	M	29 ± 6	
P00128	QCR7, 24-40 (25)	Cytochrome b-c1 complex subunit 7	M	20 ± 5	
P32628	RAD23, 364-371 (365)	UV excision repair protein RAD23	C, N, M	23 ± 9	
P39729	RBG1, 224-240 (224)	GTP-binding protein RBG1	C	7 ± 4	
Q99258	RIB3, 125-136 (133)	3,4-dihydroxy-2-butanone 4-phosphate synthase	C, M	45 ± 6	
Q99258	RIB3, 50-73 (56)	3,4-dihydroxy-2-butanone 4-phosphate synthase	C, M	12 ± 3	
Q12189	RKI1, 73-83 (75)	Ribose-5-phosphate isomerase	C, N	12 ± 5	
Q03305	RMT2, 397-408 (401)	Arginine N-methyltransferase 2	C, N	20 ± 4	
P11745	RNA1, 169-182 (180)ᵍ	Ran GTPase-activating protein 1	C, N	14 ± 3	
P49723	RNR4, 221-235 (228)	Ribonucleoside-diphosphate reductase small chain 2	C, N	12 ± 3	

Chapter 3: Oxidation-sensitive S. cerevisiae proteins   111

ID	Gene, position	Protein	N/C	Value	Ref
P28000	RPC19, 83-104 (89)	DNA-directed RNA polymerase I/III subunit RPAC2	N	29 ± 6	
P41805	RPL10, 41-69 (49)	60S ribosomal protein L10	C	15 ± 5	
P41805	RPL10, 70-82 (71)	60S ribosomal protein L10	C	21 ± 4	
Q3E757	RPL11B, 144-153 (144)	60S ribosomal protein L11-B	C	8 ± 5	
P17079	RPL12B, 131-142 (141)	60S ribosomal protein L12	C	8 ± 3	[245]
Q12672	RPL21B, 101-108 (101)	60S ribosomal protein L21-B	C	9 ± 3	
P04451	RPL23B, 121-128 (122)	60S ribosomal protein L23	C	15 ± 6	
P04451	RPL23B, 13-32 (25)	60S ribosomal protein L23	C	15 ± 7	
P40525	RPL34B, 44-60 (44,47)	60S ribosomal protein L34-B	C	43 ± 7	
P39741	RPL35B, 63-77 (67)	60S ribosomal protein L35	C	8 ± 4	
P51402	RPL37B, 15-21 (19)	60S ribosomal protein L37-B	C	19 ± 4	
P51402	RPL37B, 33-43 (34,37)	60S ribosomal protein L37-B	C	38 ± 4	
P14796	RPL40A, 13-21 (15;20)	60S ribosomal protein L40	C	66 ± 8	
P14796	RPL40A, 39-48 (39)	60S ribosomal protein L40	C	18 ± 3	
P14796	RPL40A, 8-17 (15)	60S ribosomal protein L40	C	26 ± 3	
P02405	RPL42B, 68-76 (74)	60S ribosomal protein L42	C	15 ± 3	
P49631	RPL43B, 37-45 (40,43)	60S ribosomal protein L43	C	46 ± 6	
P49626	RPL4B, 85-95 (94)	60S ribosomal protein L4-B	C	11 ± 5	
P26321	RPL5, 55-85 (62)	60S ribosomal protein L5	C	13 ± 4	
P29453	RPL8B, 150-171 (170)	60S ribosomal protein L8-B	C	9 ± 7	
P38886	RPN10, 110-127 (115)	26S proteasome regulatory subunit RPN10	N	17 ± 4	
P46654	RPS0B, 135-165 (162)	40S ribosomal protein S0-B	C	18 ± 8	[245]
P26781	RPS11B, 117-133 (128)	40S ribosomal protein S11	C	18 ± 2	
P26781	RPS11B, 58-67 (58)	40S ribosomal protein S11	C	18 ± 3	
P39516	RPS14B, 72-83 (72)	40S ribosomal protein S14-B	C	12 ± 6	
P14127	RPS17B, 34-45 (35)	40S ribosomal protein S17-B	C	8 ± 3	
P23248	RPS1B, 65-83 (69)	40S ribosomal protein S1-B	C	8 ± 2	
Q3E754	RPS21B, 16-22 (17)	40S ribosomal protein S21-B	C	20 ± 6	
Q3E7Y3	RPS22B, 72-78 (72)	40S ribosomal protein S22-B	C	13 ± 4	
P32827	RPS23B, 83-109 (92)	40S ribosomal protein S23	C	8 ± 4	
P39939	RPS26B, 67-82 (74;77)	40S ribosomal protein S26-B	C	46 ± 9	
P41057	RPS29A, 23-32 (24)	40S ribosomal protein S29-A	C	47 ± 9	
P41057	RPS29A, 33-40 (39)	40S ribosomal protein S29-A	C	46 ± 9	
P41058	RPS29B, 23-32 (24)	40S ribosomal protein S29-B	C	41 ± 7	
P41058	RPS29B, 33-40 (39)	40S ribosomal protein S29-B	C	55 ± 13	
P05759	RPS31, 44-62 (45,50)§	40S ribosomal protein S31	C	46 ± 4	
P05759	RPS37, 68-73 (68)	40S ribosomal protein S31	C	24 ± 7	
P26783	RPS5, 58-92 (87)	40S ribosomal protein S5	C	39 ± 8	
P05754	RPS8B, 165-178 (168)	40S ribosomal protein S8	C	9 ± 4	
P05754	RPS8B, 179-195 (179)	40S ribosomal protein S8	C	14 ± 2	
Q03940	RVB1, 414-431 (422)	RuvB-like protein 1	N	14 ± 9	
Q12464	RVB2, 221-232 (224)	RuvB-like protein 2	N	13 ± 6	
P39954	SAH1, 192-199 (198)§	Adenosylhomocysteinase	C	16 ± 4	

ID	Gene, position	Protein name	Loc.	Value	Refs
P39954	SAH1, 218-236 (231)[§]	Adenosylhomocysteinase	C	12 ± 2	
P10659	SAM1, 90-112 (91)	S-adenosylmethionine synthetase 1	C	7 ± 7	[245]
P10080	SBP1, 31-54 (48)	Single-stranded nucleic acid-binding protein	C, N	6 ± 4	
P21243	SCL1, 63-77 (74)	Proteasome component C7-alpha	C, M	4 ± 1	
Q00711	SDH1, 274-303 (296)	Succinate dehydrogenase flavoprotein subunit	M	31 ± 3	[327]
P47052	SDH1B, 613-632 (624)	Succinate dehydrogenase flavoprotein subunit 2	M	21 ± 4	
P29478	SEC65, 98-110 (104)	Signal recognition particle subunit SEC65	ER	10 ± 4	
P33330	SER1, 182-202 (182)	Phosphoserine aminotransferase	C	11 ± 3	
P07284	SES1, 363-379 (370/373)[§]	Seryl-tRNA synthetase	C	26 ± 4	
P07284	SES1, 397-410 (400)[§]	Seryl-tRNA synthetase	C	11 ± 2	
P07284	SES1, 411-433 (413;414)[§]	Seryl-tRNA synthetase	C	22 ± 4	
Q12118	SGT2, 24-48 (39)	Small Q-rich tetratricopeptide repeat-cont. protein 2	C	13 ± 7	
P37292	SHM1, 462-479 (464)	Serine hydroxymethyltransferase	M	12 ± 2	
P37292	SHM1, 89-106 (103)	Serine hydroxymethyltransferase	M	14 ± 2	
P37291	SHM2, 188-198 (197)	Serine hydroxymethyltransferase	C	23 ± 7	[42,245]
P37291	SHM2, 73-90 (87)	Serine hydroxymethyltransferase	C	15 ± 3	[42,245]
P34223	SHP1, 78-103 (88)	UBX domain-containing protein 1	C, N	13 ± 3	
P36024	SIS2, 2-23 (21)	Protein SIS2	C, N	18 ± 3	
P32568	SNQ2, 1097-1121 (1098)	Protein SNQ2	C, M	30 ± 7	
P00445	SOD1, 138-154 (147)	Superoxide dismutase [Cu-Zn]	C, N, M	72 ± 7	[245,270,328]
P00445	SOD1, 45-69 (58)	Superoxide dismutase [Cu-Zn]	C	69 ± 4	[245,270,328]
P10591	SSA1, 4-34 (15)	Heat shock protein SSA1	C, V, N	13 ± 3	[281,283]
P10592	SSA2, 14-34 (15)	Heat shock protein SSA2	C, V, M	17 ± 5	[51,245,281,283]
P10592	SSA2, 304-313 (308)	Heat shock protein SSA2	C, M	7 ± 2	[51,245,281,283]
P11484	SSB1, 431-451 (435)	Heat shock protein SSB1	C	14 ± 3	[37,245,246]
P11484	SSB1, 431-455 (435;454)	Heat shock protein SSB1	C	19 ± 3	[37,245,246]
P40150	SSB2, 452-466 (454)	Heat shock protein SSB2	C	11 ± 3	
P32589	SSE1, 223-235 (228)	Heat shock protein homolog SSE1	C	10 ± 3	
P32590	SSE2, 375-388 (380)	Heat shock protein homolog SSE2	C	9 ± 4	
P38353	SSH1, 70-77 (74)	Sec61 protein homolog	ER	7 ± 5	
P39926	SSO2, 108-123 (122)	Protein SSO2	C, ER	39 ± 4	
P38788	SSZ1, 65-85 (81)	Ribosome-associated complex subunit SSZ1	C	29 ± 8	
Q07478	SUB2, 359-368 (360)	ATP-dependent RNA helicase SUB2	N	10 ± 5	
P53616	SUN4, 233-246 (239)	Septation protein SUN4	C, M	5 ± 2	
P00359/P00358	TDH2/3, 144-160 (150;154)*	Glyceraldehyde-3-phosphate dehydrogenase 2/3	C, M	26 ± 3	[37,42,262,263,281,283]
P02994	TEF1, 101-134 (111)	Elongation factor 1-alpha	C, M	14 ± 3	[51,245,246,263]
P02994	TEF1, 21-37 (31)	Elongation factor 1-alpha	C, M	18 ± 3	[51,245,246,263]
P02994	TEF1, 320-328 (324)	Elongation factor 1-alpha	C, M	11 ± 4	[51,245,246,263]
P02994	TEF1, 356-369 (361;368)	Elongation factor 1-alpha	C	17 ± 3	[51,245,281,283]
P02994	TEF1, 409-421 (409)	Elongation factor 1-alpha	C, M	13 ± 3	[51,245,246,263]
P16521	TEF3, 665-696 (691)	Elongation factor 3A	R	11 ± 2	
P16521	TEF3, 708-732 (731)	Elongation factor 3A	R	9 ± 4	
P16521	TEF3, 874-893 (879)	Elongation factor 3A	R	10 ± 3	
P36008	TEF4, 125-143 (129)	Elongation factor 1-gamma 2	C, R, M	7 ± 2	[37]

Chapter 3: Oxidation-sensitive *S. cerevisiae* proteins

Accession	Protein, position (length)	Description	Localization	Value	Ref
P17255	TFP1, 522-536 (532)	V-type proton ATPase catalytic subunit A	V	14 ± 3	
P17255	TFP1, 356-367 (358)	V-type proton ATPase catalytic subunit A	V	9 ± 3	
P17423	THR1, 216-231 (216)	Homoserine kinase	unknown	8 ± 3	
P16120	THR4, 416-432 (428)	Threonine synthase	C, N	9 ± 3	
P04801	THS1, 264-278 (268)	Threonyl-tRNA synthetase	C, M	26 ± 4	
P38912	TIF11, 83-101 (89)	Eukaryotic translation initiation factor 1A	R	12 ± 3	
P10081	TIF2, 121-138 (124)	ATP-dependent RNA helicase eIF4A	C, R	11 ± 5	
P87108	TIM10, 44-64 (44,61)	Mit. import inner membrane translocase subunit TIM10	M	93 ± 5	
Q01852	TIM44, 356-370 (369)	Mit. import inner membrane translocase subunit TIM44	M	10 ± 2	
P23644	TOM40, 324-330 (326)	Mitochondrial import receptor subunit TOM40	C	17 ± 3	
P07213	TOM70, 425-443 (439)	Mitochondrial import receptor subunit TOM70	M	12 ± 4	
P00942	TPI1, 115-134 (126)	Triosephosphate isomerase	C, M	7 ± 4	[51,245,246,281,327]
P00942	TPI1, 27-55 (41)	Triosephosphate isomerase	C, M	7 ± 4	[51,245,246,281,327]
Q00764	TPS1, 206-222 (210)	Trehalose-phosphate synthase	C	6 ± 4	
P15565	TRM1, 333-343 (333)	tRNA(guanine-26,N(2)-N(2)) methyltransferase	N, M	65 ± 13	
P00899	TRP2, 121-135 (133)	Anthranilate synthase component 1	C	6 ± 4	
P00931	TRP5, 306-323 (321)	Tryptophan synthase	C, N	12 ± 2	
P00931	TRP5, 402-422 (420)	Tryptophan synthase	C, N	7 ± 3	
P29509	TRR1, 138-153 (142;145)	Thioredoxin reductase 1	C	39 ± 6	[37,245]
P02992	TUF1, 192-218 (213)	Elongation factor Tu	M	8 ± 2	[327]
P25515	UBA1, 140-150 (144)	Ubiquitin-activating enzyme E1 1	C, N	7 ± 4	
P25515	UBA1, 596-603 (600)	Ubiquitin-activating enzyme E1 1	C, N	27 ± 3	
P15731	UBC4, 103-129 (108)	Ubiquitin-conjugating enzyme E2 4	N	19 ± 4	[245,246]
P32861	UGP1, 351-369 (364)	UTP–glucose-1-phosphate uridylyltransferase	C, M	11 ± 3	
P07259	URA2, 1832-1854 (1832)	Protein URA1	C, M	11 ± 4	
P07259	URA2, 325-337 (330)	Protein URA2	C	21 ± 3	
P13298	URA5, 11-23 (19)	Orotate phosphoribosyltransferase 1	C, N	5 ± 3	
P15700	URA6, 124-138 (130)	Uridylate kinase	C, N	6 ± 4	
P28274	URA7, 204-215 (214)	CTP synthase 1	C	14 ± 5	
P39968	VAC8, 300-317 (315)	Vacuolar protein 8	V	18 ± 4	
P39968	VAC8, 341-359 (352)	Vacuolar protein 8	V	23 ± 5	
P39968	VAC8, 428-447 (445)	Vacuolar protein 8	V	12 ± 3	
P07806	VAS1, 765-775 (768)	Valyl-tRNA synthetase	C, M	16 ± 5	
P07806	VAS1, 998-1031 (1015)	Valyl-tRNA synthetase	C, M	15 ± 7	
P41807	VMA13, 76-94 (92)	V-type proton ATPase subunit H	V	14 ± 3	
P16140	VMA2, 170-189 (188)	V-type proton ATPase subunit B	C, V	17 ± 3	
P32366	VMA6, 322-335 (329)	V-type proton ATPase subunit d	V	13 ± 4	
P32563	VPH1, 386-403 (394,396)	V-type proton ATPase subunit a	V	11 ± 3	
P47075	VTC4, 612-622 (614)	Vacuolar transporter chaperone 4	V	24 ± 5	
P25654	YCR090C, 112-125 (124)	UPF0587 protein YCR090C	C, N	34 ± 8	
P25491	YDJ1, 140-149 (143;146)	Mitochondrial protein import protein MAS5	M	64 ± 5	
P25491	YDJ1, 181-200 (185;188)	Mitochondrial protein import protein MAS5	M	46 ± 4	
P25491	YDJ1, 364-382 (370)	Mitochondrial protein import protein MAS5	C	7 ± 2	

Q03034	YDR539W, 58-74 (69)	Uncharacterized protein YDR539W	C	15 ± 4	
P53111	YGL157W, 232-247 (240)	Putative uncharacterized oxidoreductase YGL157W	C, N	31 ± 6	
P39676	YHB1, 320-333 (328)	Flavohemoprotein	C, M	13 ± 4	
P38708	YHR020W, 387-405 (396)	Putative prolyl-tRNA synthetase YHR020W	C	9 ± 4	
P52553	YKE2, 72-84 (82)	Prefoldin subunit 6	C	88 ± 10	
P36015	YKT6, 72-96 (76)	Synaptobrevin homolog YKT6	C, G, V, M	13 ± 3	
Q06252	YLR179C, 137-146 (142)	Uncharacterized protein YLR179C	C, N	4 ± 1	
P32795	YME1, 600-627 (616)	Protein YME1	M	2 ± 1	
P36010	YNK1, 116-129 (118)	Nucleoside diphosphate kinase	C, M	6 ± 3	[326-328]
P53981	YNL010W, 167-188 (177)	Uncharacterized phosphatase YNL010W	C, N	10 ± 5	
Q12305	YOR285W, 93-103 (98)	Putative thiosulfate sulfurtransferase YOR285W	ER, M	21 ± 4	
Q12414	YPL257W-B, 25-45 (43)	Transposon Ty1-PL Gag-Pol polyprotein	unknown	10 ± 3	
P41920	YRB1, 134-154 (135)	Ran-specific GTPase-activating protein 1	C, N	16 ± 6	
Q12512	ZPS1, 123-142 (123;130)	Protein ZPS1	C, V	96 ± 4	[257]
P32527	ZUO1, 162-178 (167)	Zuotin	C, M	13 ± 6	

Chapter 3: Oxidation-sensitive *S. cerevisiae* proteins   115

Table 3.B. H_2O_2-sensitive yeast proteins. *Exponentially growing S. cerevisiae cells were treated with 0.5 mM H_2O_2 for 15 min and then subjected to the OxICAT procedure. Peptides are grouped according to i) their function and ii) gene name. Gene name, protein name, and Swiss-Prot accession numbers are based on annotations in SGD. "% oxidation before stress" and "% oxidation after stress" denotes the average oxidation of the redox-sensitive cysteine (second column, in brackets), obtained from at least three independent experiments ± standard deviation. Only proteins with at least a 1.4-fold change between untreated and H_2O_2-treated yeast cells are listed. For references, refer to Appendix, Table 3.A. $: essential gene; *: peptide identical in both isoenzymes; ^{a}: at this mass, two different peptides were identified.*

Cellular process	Swiss-Prot	Gene, Peptide (Cys)	Protein name	Cellular compartment	% oxidation before stress	% oxidation after stress	Fold change
Protein translation							
	Q06053	DUS3, 210-225 (224)	tRNA-dihydrouridine synthase 3	C, N	13 ± 2	23 ± 2	1.7
	P32324	EFT2, 121-144 (136)	Elongation factor 2	C	11 ± 2	20 ± 3	1.8
	P09436	ILS1, 99-120 (119)$	Isoleucyl-tRNA synthetase	C	25 ± 5	38 ± 7	1.6
	P15180	KRS1, 481-496 (486)$	Lysyl-tRNA synthetase	C	13 ± 2	23 ± 4	1.8
	P11745	RNA1, 169-182 (180)$	Ran GTPase-activating protein 1	C, N	14 ± 3	47 ± 10	3.3
	P17079	RPL12B, 131-142 (141)	60S ribosomal protein L12	C	8 ± 3	16 ± 2	1.9
	Q12672	RPL21B, 101-108 (101)	60S ribosomal protein L21-B	C	9 ± 3	21 ± 2	2.4
	P51402	RPL37B, 33-43 (34;37)	60S ribosomal protein L37-B	C	38 ± 4	54 ± 8	1.4
	P14796	RPL40A, 39-48 (39)	60S ribosomal protein L40	C	18 ± 3	36 ± 6	2.0
	P02405	RPL42B, 68-76 (74)	60S ribosomal protein L42	C	15 ± 3	39 ± 3	2.7
	P26781	RPS11B, 117-133 (128)	40S ribosomal protein S11	C	18 ± 2	27 ± 5	1.5
	P05759	RPS31, 44-62 (45;55)$	40S ribosomal protein S31	C	46 ± 4	67 ± 4	1.4
	P05754	RPS8B, 179-195 (179)	40S ribosomal protein S8	C	14 ± 2	23 ± 4	1.7
	P07284	SES1, 363-379 (370;373)$	Seryl-tRNA synthetase	C	26 ± 4	42 ± 6	1.6
	P07284	SES1, 397-410 (400)$	Seryl-tRNA synthetase	C	11 ± 2	17 ± 3	1.5
	P02994	TEF1, 320-328 (324)	Elongation factor 1-alpha	C, M	11 ± 4	20 ± 4	1.8
	P02994	TEF1, 356-369 (361;368)	Elongation factor 1-alpha	C	17 ± 3	24 ± 2	1.4
	P02994	TEF1, 409-421 (409)	Elongation factor 1-alpha	C, M	13 ± 3	26 ± 5	2.0
	P36008	TEF4, 125-143 (129)	Elongation factor 1-gamma 2	C, R, M	7 ± 2	12 ± 1	1.8
Amino acid metabolism							
	P28777	ARO2, 210-224 (221)	Chorismate synthase	C	13 ± 3	29 ± 5	2.3
	P32449	ARO4, 69-90 (76)	Phospho-2-dehydro-3-deoxyheptonate aldolase	C, N	9 ± 2	26 ± 5	3.0
	P49089/P49090	ASN1/2, 396-406 (404)*	Asparagine synthetase 1/2	C	7 ± 2	14 ± 2	2.2
	P13663	HOM2, 250-263 (259)	Aspartate-semialdehyde dehydrogenase	C, N	21 ± 3	31 ± 3	1.4
	P10869	HOM3, 38-56 (53)	Aspartokinase	C	5 ± 1	10 ± 1	2.0
	P25605	ILV6, 261-272 (270)	Acetolactate synthase small subunit	M	9 ± 4	14 ± 4	1.5
	P39954	SAH1, 218-236 (231)$	Adenosylhomocysteinase	C	12 ± 2	22 ± 5	1.8
	P37292	SHM1, 89-106 (103)	Serine hydroxymethyltransferase	M	14 ± 2	20 ± 3	1.4

Accession	Gene, range	Protein name	Location			
P17423	THR1, 216-231 (216)	Homoserine kinase	unknown	8 ± 3	13 ± 4	1.6
P00931	TRP5, 402-422 (420)	Tryptophan synthase	C, N	7 ± 3	27 ± 4	4.0
Carbohydrate and energy metabolism						
P19414	ACO1, 376-392 (382)	Aconitase 1	C, M	7 ± 2	15 ± 2	2.0
P00330	ADH1, 277-299 (277,278)	Alcohol dehydrogenase 1	C	17 ± 5	40 ± 5	2.3
P14540	FBA1, 96-115 (112)[§]	Fructose-bisphosphate aldolase	C, M	13 ± 2	23 ± 4	1.8
P09624	LPD1, 45-58 (54)	Dihydrolipoyl dehydrogenase	M	12 ± 2	27 ± 2	2.3
P06169	PDC1, 213-224 (221;222)	Pyruvate decarboxylase isozyme 1	C, N	22 ± 4	33 ± 6	1.5
P00560	PGK1, 83-108 (98)[§]	Phosphoglycerate kinase	C, M	9 ± 3	15 ± 3	1.7
P00359/P00358	TDH2/3, 144-160 (150:154)*	Glyceraldehyde-3-phosphate dehydrogenase 2/3	C, M	26 ± 3	75 ± 5	3.0
Protein folding and stress response						
P38013	AHP1, 114-124 (120)	Peroxiredoxin type-2	C	6 ± 2	19 ± 3	3.1
P38013	AHP1, 49-79 (62)	Peroxiredoxin type-2	C	14 ± 2	51 ± 2	3.6
P39077	CCT3, 512-531 (518)[§]	T-complex protein 1 subunit gamma	C	14 ± 3	28 ± 5	1.9
P42943	CCT7, 452-469 (454)[§]	T-complex protein 1 subunit eta	C	11 ± 3	28 ± 5	2.4
P53691	CPR6, 150-165 (156)	Peptidyl-prolyl cis-trans isomerase CPR6	C	7 ± 2	13 ± 2	1.8
P40150	SSB2, 452-466 (454)	Heat shock protein SSB2	C	11 ± 3	17 ± 2	1.6
P25491	YDJ1, 181-200 (185;188)	Mitochondrial protein import protein MAS5	M	46 ± 4	69 ± 6	1.5
P25491	YDJ1, 364-382 (370)	Mitochondrial protein import protein MAS5	C	7 ± 2	11 ± 3	1.7
Lipid metabolism						
P41338	ERG10, 343-362 (358)[a,§]	Acetyl-CoA acetyltransferase	C	12 ± 3	28 ± 3	2.3
P29704	ERG9, 340-350 (341)[§]	Squalene synthetase	ER, M	9 ± 2	19 ± 3	2.1
P11986	INO1, 19-29 (19)	Inositol-1-phosphate synthase	C	34 ± 4	54 ± 9	1.6
Protein degradation						
P25694	CDC48, 104-119 (115)[§]	Cell division control protein 48	ER, N	12 ± 3	18 ± 3	1.5
P34223	SHP1, 78-103 (88)	UBX domain-containing protein 1	C, N	13 ± 3	20 ± 2	1.6
Other metabolisms						
Q04894	ADH6, 141-174 (163)	NADP-dependent alcohol dehydrogenase 6	C	24 ± 3	40 ± 6	1.7
P16550	APA1, 292-321 (306)	AP4A phosphorylase	C, N	13 ± 2	21 ± 4	1.7
P38625	GUA1, 438-455 (450)	GMP synthase	unknown	9 ± 3	18 ± 4	2.1
Miscellaneous						
P14832	CPR1, 36-53 (38)[a]	Peptidyl-prolyl cis-trans isomerase CPR1	C	12 ± 3	28 ± 3	2.3
P32614	FRDS1, 446-459 (454)	Probable fumarate reductase	C, M	6 ± 2	20 ± 3	3.4
P18562	FUR1, 195-216 (215)	Uracil phosphoribosyltransferase	unknown	11 ± 6	19 ± 4	1.7
P53091	MCM6, 813-832 (815)[§]	DNA replication licensing factor MCM6	C, N	20 ± 5	45 ± 7	2.3
P15646	NOP1, 273-288 (275)[§]	rRNA 2'-O-methyltransferase fibrillarin	C, N	9 ± 3	15 ± 2	1.8
P39926	SSO2, 108-123 (122)	Protein SSO2	C, ER	39 ± 4	65 ± 13	1.7
P17255	TFP1, 356-367 (358)	V-type proton ATPase catalytic subunit A	V	9 ± 3	18 ± 2	2.0
Q12305	YOR285W, 93-103 (98)	Putative thiosulfate sulfurtransferase YOR285W	ER, M	21 ± 4	49 ± 5	2.4

Table 3.C. Superoxide-sensitive yeast proteins. *Exponentially growing S. cerevisiae cells were treated with 3 mM paraquat (superoxide-generator) for 15 min and then subjected to the OxICAT procedure. Peptides are grouped by i) their function and ii) gene name. Gene name, protein name, and Swiss-Prot accession numbers are based on annotations in SGD. "% oxidation before stress" and "% oxidation after stress" denotes the average oxidation of the redox-sensitive cysteine (second column, in brackets), obtained from at least three independent experiments ± standard deviation. Only proteins with at least a 1.4-fold change between untreated and superoxide-treated yeast cells are listed. For references, refer to Appendix, Table 3.A.*

Cellular process	Swiss-Prot	Gene, Peptide (Cys)	Protein name	Cellular compartment	% oxidation before stress	% oxidation after stress	Fold change
Protein translation							
	P23248	RPS1B, 65-83 (69)	40S ribosomal protein S1-B	C	8 ± 2	13 ± 1	1.6
Amino acid metabolism							
	P10869	HOM3, 38-56 (53)	Aspartokinase	C	5 ± 1	20 ± 4	3.6
	P00931	TRP5, 306-323 (321)	Tryptophan synthase	C, N	12 ± 2	20 ± 4	1.6
	P07259	URA2, 325-337 (330)	Protein URA2	C	21 ± 3	34 ± 3	1.6
Carbohydrate and energy metabolism							
	P19414	ACO1, 376-392 (382)	Aconitase 1	C, M	7 ± 2	21 ± 3	2.9
Protein folding and stress response							
	P38720	GND1, 287-295 (287)	6-phosphogluconate dehydrogenase	C, M	15 ± 4	32 ± 4	2.1
	P10592	SSA2, 304-313 (308)	Heat shock protein SSA2	C, M	7 ± 2	18 ± 2	2.5
Other metabolisms							
	P09440	MIS1, 216-229 (224)	C-1-tetrahydrofolate synthase	M	10 ± 3	22 ± 2	2.1

4 Collapse of the Cellular Redox Balance: A Key Event for Chronological Aging in Yeast?

4.1 Introduction

All living animals undergo a physiological decline with age. Despite decades of intensive studies, no consensus has yet emerged regarding the primary cause of aging. One leading theory is the "free radical theory of aging", which states that the progressive decline in the functional capacity of aging organisms is the result of accumulating oxidative damage caused by reactive oxygen species (ROS) [330]. In support of this theory, several studies reported the correlation between an increase in ROS levels and aging cells and organisms [263,331-334]. ROS, such as hydrogen peroxide (H_2O_2), superoxide anions ($\cdot O_2^-$), and hydroxyl radicals ($\cdot OH$), are constantly generated as by-products of oxidative phosphorylation and NADPH oxidases [5,330]. With the use of various antioxidant compounds, ROS scavengers, and reducing systems, such as the thioredoxin and glutaredoxin systems, cells spent large amounts of their energy to offset elevated ROS production in an effort to maintain cellular redox homeostasis [5,335]. Accumulation of ROS beyond the capacity of the antioxidant machinery and cellular reducing systems leads to DNA, lipid, and protein damage and causes oxidative stress conditions that are often deleterious to cells and organisms [8,18,19].

Based on the high and specific reactivity of certain protein cysteine residues towards oxidation, the oxidation status of thiol groups can be used to assess the redox status of cells [30]. Cysteine residues exist either fully reduced as thiols (R-SH) or thiolate anions (R-S$^-$), or reversibly oxidized as sulfenic acids (R-SOH), disulfide bonds (R-S-S-R) or mixed disulfide bonds with glutathione (R-S-SG). Presence of high concentrations of oxidants can also lead to the irreversible oxidation of cysteines to sulfinic (R-SO$_2$H) or sulfonic (R-SO$_3$H) acids [314]. Cysteine oxidation has the potential to alter the structure, stability and function of a protein, and it is not surprising that the group of proteins that use redox-active cysteines to regulate their activity expanded rapidly over the past few years [25-29]. Oxidative thiol modifications in redox-sensitive proteins have been shown to play important roles in many cellular processes, including transcription, metabolism, stress protection, apoptosis, and signal transduction [38,70,225,336].

Accumulation of oxidatively modified proteins is a common feature of cell senescence in most cell types [8,51,245,281] and is linked to a variety of diseases such as cancer, diabetes or neurodegenerative disorders [281,331,337]. Moreover, many studies have shown an inverse correlation between the level of protein oxidation and the age of an organism, and revealed that increased antioxidant capacity often results in life span extension [263,338-340]. Even though these studies support the free radical theory of aging, it is still unclear whether the accrual of oxidatively damaged proteins is the cause of aging or simply a consequence or byproduct of the aging process.

Saccharomyces cerevisiae with its well defined genome, numerous human orthologs, and large variety of available biochemical assays and genetic tools, has long been used as a model system to study oxidative stress and aging [263,268,330,338,341]. Two different types of aging have been described in *S. cerevisiae*, replicative and chronological aging. Replicative aging refers to the mitotic capacity of yeast cells and is defined by the number of daughter cells produced by a mother cell prior to senescence [332]. In contrast, chronological aging, which models aging of post-mitotic mammalian cells (*e.g.*, neurons), is defined as the length of time stationary yeast cells remain viable in a non-dividing, high metabolic state [342,343]. In support of the free radical theory of aging, it has been shown that the chronological life span (CLS) decreases in strains lacking superoxide dismutase or catalase [312,343,344], while addition of the physiological antioxidant glutathione or overexpression of superoxide dismutase (SOD) extends CLS [339,343,345-347].

Throughout their chronological life span, *S. cerevisiae* cells use different carbon sources for energy metabolism. When grown on a fermentable substrate like glucose, yeast cells first generate most of their energy by glycolysis [268]. After depletion of these fermentable carbons, yeast cells undergo a diauxic shift to a respiratory-based metabolism, which involves the activation of the mitochondrial oxidative phosphorylation machinery [3,4]. In this post-diauxic phase, growth continues and cells use ethanol and acetate, the fermentative breakdown products of glucose for their energy metabolism. Once all extracellular nutrients are internalized, yeast cells switch into a non-dividing, stationary phase [265]. Prior to entry and in stationary phase, they actively respond to changes in nutrient availability by undertaking various genetically controlled physiological adjustments to enhance survival. They maintain high metabolic rates and prevent starvation as they produce energy mainly by catabolizing the accumulated reserve metabolites [268,340].

Caloric restriction (CR) defined by a reduction in nutrient availability without malnutrition, has been demonstrated to increase the CLS of yeast and the life span of many different organisms, including nematodes, rodents and fruit flies [348-351]. An increase of CLS through CR is accomplished by lowering the glucose concentration from 2% to 0.5% or less in the initial culture medium [342,343,352,353]. Alternatively, yeast are grown in glucose medium for several days and switched to water (extreme caloric restriction) [342,343,352,353]. In addition, CLS can also be extended by growing yeast on a non-fermentable carbon source such as glycerol or by increasing the osmolarity of the medium [352-355].

In this study, we used a quantitative proteomic approach called OxICAT to precisely determine the extent and onset of oxidative stress in chronologically aging *S. cerevisiae*. Our results demonstrate that yeast cells undergo a global collapse of the cellular redox homeostasis, which precedes cell death. The onset of this collapse appears to correlate to the yeast life span, as caloric restriction increases the life span and delays the redox collapse. We identified a subset of proteins, including thioredoxin reductase and Cdc48, whose oxidative thiol modifications substantially precede the general redox collapse in both non-restricted and restricted glucose media. These early targets of oxidation might contribute to the altered physiology of chronologically aging yeast.

4.2 Material and Methods

4.2.1 Strains and cell growth

S. cerevisiae strain EG103 (DBY746; *MATα, leu2-3 112 his3Δ1 trp1-289a ura3-52*) was grown aerobically at 30°C in synthetic complete (SC) medium (0.67% yeast nitrogen base without amino acids) supplemented with complete amino acid dropout solution and either 2% glucose (synthetic complete dextrose; herein referred to as "SCD media") or 0.5% glucose (caloric restriction; herein referred to as "CR media") as carbon source. Cells were picked from fresh colonies and grown twice to late logarithmic phase in SCD or CR media to synchronize the cultures in an early, healthy growth-phase. Cells were then diluted into fresh media using flasks with volume-to-medium ratio of 5:1. Growth was monitored by measuring the optical density at OD_{600}. For extreme caloric restriction, yeast cells growing in SCD or CR media for two days were harvested by centrifugation. Then, cells were washed three times with $_{dd}H_2O$, resuspended in $_{dd}H_2O$ and continued to incubate at 30°C with vigorous shaking.

4.2.2 Determination of cell viability

The viable cell count was determined by propidium iodide (PI) staining [356]. Chronologically aging yeast cells were harvested by centrifugation and the pellet was resuspended in 3.5 µg/ml PI in phosphate buffered saline (PBS). Cells were incubated on ice for 10 min and 400-600 individual cells per sample were analyzed with a Nikon Eclipse E600 microscope with G-2A filter (excitation BP 510-560, emission LP 590 nm) for positive PI staining, indicating dead cells. Cell viability was determined based on PI staining and expressed as mean ± standard deviation of at least three independent experiments. Cell viability of the yeast culture in logarithmic phase was set to 100%.

4.2.3 Differential thiol trapping of chronological aging yeast

Yeast cells were grown in SCD or CR medium at 30°C with continuous shaking. At midlogarithmic phase (OD_{600} of 0.5), 12 ml of cell culture, and from here on every 24 hours 4 ml of cell culture, was harvested onto trichloroacetic acid (TCA; 10% (w/v) final concentration) to stop all thiol-disulfide exchange reactions. The TCA-precipitated samples were incubated on ice for 30 min and the thiol trapping protocol established by Leichert *et al.* [49] and modified by Brandes *et al.* (see Chapter 3.2.4 – 3.2.7 for detailed OxICAT protocol) was applied. OxICAT samples of chronological aging yeast cells in SCD media, CR media or after switch into water were harvested for 4, 7, and 10 days, respectively.

4.2.4 Clustering analysis

The open-source software TIGR MultiExperimentViewer v4.4 (MEV; http://www.tm4.org/mev/) [357] and the algorithm *k-means clustering with Euclidean distance* (implemented in MEV) were used for clustering analysis of peptides from yeast cultivated in 2% and 0.5% glucose media. Input for clustering were peptide lists with the corresponding oxidation levels defined by OxICAT. Values missing in those lists as a result of insufficient MS quantification were predicted with *Coupled Two Way Clustering* (CTWC) (Weizmann Institute of Science, Israel; http://ctwc.weizmann.ac.il/) [358]. The prediction was based on 5 neighbors when more than 30% of the values were known.

4.2.5 Prediction of fold propensity

Fold predictions to estimate the probability of a cysteine-containing amino acid stretch to fold were performed with FoldIndex (http://bip.weizmann.ac.il/ fldbin/findex) [254]. For FoldIndex, default settings were chosen and the amino acid sequence of all cysteine-containing proteins that were identified in chronological yeast aging with OxICAT served as input.

4.2.6 Analysis of glucose concentration

Glucose concentration in the growth media was analyzed using an enzymatic assay kit (Glucose (HK) Assay Kit, Sigma-Aldrich) that combines hexokinase and glucose-6-phosphate dehydrogenase activities [247]. For the measurement, the yeast cell culture was diluted with water to 0.05 - 5 mg of glucose/ml. Then, 10 - 200 µl of this solution, corresponding to 0.5 - 50 µg of glucose, was added to 1 ml of "Glucose assay reagent" (prepared according to manufacturer's protocol), mixed and incubated for 15 min at 25°C. Then, the absorbance was determined at 340 nm using a Cary 3 UV spectrophotometer (Varian Instruments). Preparation of blanks and calculation of glucose concentration were carried out according to the manufacturer's protocol.

4.2.7 Determination of total cellular ATP levels

To determine total cellular ATP levels, 1 ml of aging yeast culture ($\sim$1-4 x 10^7 cells) in SCD or CR media was harvested by centrifugation and washed three times with $_{dd}H_2O$. Then,

200 µl of boiling $_{dd}H_2O$ was added to the pellet. Cells were vortexed and immediately heated for 5 min in boiling water [359]. Lysed cells were cooled on ice for 2 min, centrifuged (12,000 x g, 2 min, 4°C) and 50 µl of the supernatant was mixed with 50 µl of luciferase assay buffer (200 mM KH_2PO_4 pH 7.5, 50 mM glycylglycine, 0.2 µM luciferase, 0.14 µM luciferin, 0.4 mM EDTA, 0.1 mg/ml bovine serum albumin) for chemiluminescence measurement [360] in a POLARstar Omega microplate reader (BMG Labtech). Results were normalized by cell number and expressed as mean ± standard deviation of three independent experiments.

4.2.8 Determination of intracellular GSH, GSSG, and cysteine concentration and calculation of redox potentials

For determination of intracellular GSH, GSSG and cysteine concentrations, 1 ml of aging yeast culture (~1-4 x 10^7 cells) in SCD or CR media was harvested by centrifugation and washed three times with ice-cold PBS. 150 µl of PBS was added to the cell pellet and yeast cells were lysed by vigorously vortexing them in the presence of glass-beads. Next, an equal volume of metaphosphoric acid solution (16.8 mg/ml HPO_3, 2 mg/ml EDTA, 9 mg/ml NaCl) was added to the cell suspension and briefly vortexed. After the precipitated proteins were sedimented by centrifugation (13,000 x g, 10 min, 4°C), the supernatant was transferred into a new tube and thiols were alkylated with monoiodoacetic acid (7 mM final concentration in water). After vortexing and adjustment of the pH to ~8 with saturated K_2CO_3, the reaction was allowed to proceed for ~1 h in the dark at room temperature. Finally, an equal volume of freshly prepared 2,4-dinitrofluorobenzene solution in 1.5% (v/v) absolute ethanol was added to the mixture, which was vortexed and incubated for at least 4 h in the dark at room temperature.

The following steps were conducted by Dr. Victor Vitvitsky in the Banerjee Lab at the University of Michigan. The N-dinitrophenyl derivatives of cysteine, GSH and GSSG were separated by HPLC on a Waters µ Bondapak NH_2 column (300 mm x 3.9 mm, 10 µm) at a flow rate of 1 ml/min. The mobile phase consisted of solvent A (4:1 methanol/water mixture) and solvent B (154 g ammonium acetate in 122 ml of water and 378 ml of glacial acetic acid; 500 ml of the resulting solution are added to 1000 ml of solvent A). For elution, the following conditions were used: from 0-10 min isocratic 25% solvent B, from 10-30 min linear gradient 25-100% solvent B, from 30-34 min 100% solvent B, from 34-36 min 100-25% solvent B and finally 36-45 min 25% solvent B. Prior to each injection, the column was equilibrated with 25% solution B for 10 min. Elution of the metabolites was monitored by their absorbance at 355 nm. Under these conditions, cysteine, GSH, and GSSG exhibited retention times of 18, 26, and 29 min, respectively. The concentration of the individual sulfur-containing compounds was determined by comparing the integrated peak areas with independently generated calibration curves for each compound.

Redox potentials were calculated using the determined intracellular concentration of GSH and GSSG and the Nernst equation (equation 1)

equation 1 $$E^h = E^0 + 2{,}303 \left(\frac{kT}{nF}\right) \cdot \log\frac{[GSSG]}{[GSH]^2}$$

where E^0 is the standard electrode potential (-240 mV) for reduced glutathione at pH 7, k is the Boltzmann's constant (8.31 J mol^{-1} K^{-1}), T is the temperature (298 K), n is the number of electrons transferred (2 electrons) and F is the Faraday constant (96,406 J V^{-1}) [361]. Results were normalized by cell number and expressed as mean ± standard deviation of three independent experiments.

4.3 Results and Discussion

4.3.1 Glucose concentration — A determinant of chronological life span in yeast

A well-adapted model system to study aging of post-mitotic cells is the analysis of the chronological life span (CLS) of *S. cerevisiae*. In CLS assays, the length of time that a population of non-dividing cells remains viable under stationary growth conditions is monitored [342,343]. In the past, several growth conditions have been described that extend CLS relative to the chronological life span in standard culture conditions of 2% glucose. These conditions include cultivation in non-fermentable carbon sources (*e.g.*, 3% glycerol), caloric restriction by reducing the initial glucose concentration in the media to 0.5% glucose or less, or switching stationary phase yeast cultures seeded and cultivated for two days in 2% glucose media into water [263,334,362].

We measured the chronological life span of the *EG103* yeast strain that was cultivated continuously in 2% glucose SCD media or 0.5% glucose CR media, and after transferring *EG103* cells into water after two days of cultivation in SCD or CR media. Life span measurements were started by diluting overnight cultures into fresh SCD or CR media to an initial density of 5×10^5 cells/ml (OD_{600} of ~0.05). At midlogarithmic phase (OD_{600} of 0.5), and from here on every 24 hours, aliquots were removed and the survival of the yeast cultures was evaluated under the microscope using propidium iodide, which stains all dead cells in a culture bright red. Cell viability of the yeast population was calculated and described relative to the midlogarithmic phase culture, which was set to 100% viability. In SCD media, more than 80% of the cells in culture were viable after four days of cultivation (Fig. 4.1). Then, however, the viability of the yeast culture dropped continuously over the next few days, resulting in a mean life span of ~7 days. In CR media, the yeast population lived significantly longer as compared to SCD media. About 80% of the population was still viable after seven days and the mean life span in CR media was ~11 days, an increase of ~60% relative to CLS in 2% glucose (Fig. 4.1). These differences in viability are in line with earlier reports [263,281,363] and reflect the importance of the initial glucose concentration for the CLS of yeast cells.

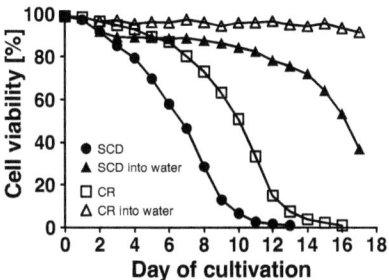

Figure 4.1. Cell viability [%] of a *S. cerevisiae* population cultivated in different media conditions. *Yeast cells were cultivated in 2% glucose SCD media (●), 0.5% glucose CR media (□), or shifted into water after two days of cultivation in SCD (▲) or CR media (Δ). Cell viability was determined by propidium iodide, which stains all dead cells in a culture bright red.*

Not only the initial glucose concentration influences the CLS but also the composition of the media that yeast cells are cultivated in once cell division has ceased. As observed before [343], cells grown in SCD or CR media for two days and then shifted into water displayed mean life spans that greatly exceed CLS of yeast populations continuously cultivated in SCD or CR media (Fig. 4.1). These results suggest that a decrease in metabolic rates, as observed under these cultivation conditions [268], extends life span, a conclusion that supports the free radical theory of aging. It has also been discussed that substances secreted by yeast cells into the media during the early phases of cultivation might shorten the yeast's life span. It has been shown that CLS is increased by reducing the extracellular acetic acid concentration, whose accumulation in the culture media as by-product of fermentative metabolism correlates with the initial glucose concentration [363]. These results led to the conclusion that yeast populations might promote extended life span by either reducing acetic acid accumulation or increasing resistance to acetic acid-induced death [363]. Intracellular acidification might also be used in a yeast culture as signal to indicate high population density and limited nutrients, which ultimately activates apoptosis [364].

4.3.2 Detection of oxidation-sensitive protein cysteines during chronological aging

In support of the free radical theory of aging, previous studies suggested that ROS accumulate during chronological yeast aging and cause oxidation of cellular macromolecules, such as proteins [8,51,245,281]. However, largely due to a lack of appropriate methodologies, none of these studies were able to quantitatively assess the extent, determine the precise onset, or evaluate the physiological consequences of oxidative stress in aging yeast or any other organism [263,281]. With the OxICAT method (for detailed description, see Chapter 3.3.1 and Fig. 3.2) [49], we have now established a highly quantitative measure of protein oxidation that focuses on the redox status of cysteine thiols. Identification of the affected proteins and determination of their degree of oxidative thiol modification should provide us not only with a quantitative read-out about the extent of oxidative stress *in vivo*, but supply us with valuable information about how changes in the oxidation status might affect activity and regulation of the respective proteins. This information should allow us to explain some of the physiological alterations that are observed in aging cells and organisms.

4.3.3 Chronological aging in yeast — Collapse of the cellular redox homeostasis?

To assess the extent of oxidative stress that yeast cells are exposed to during chronological aging, the thiol redox status of yeast proteins using the OxICAT technique [49] was analyzed. *S. cerevisiae* was cultivated in 2% glucose SCD media and samples were removed for OxICAT analysis in exponential phase and subsequently every 24 hours for the first 4 days after the inoculation. During this cultivation period, less than ~25% of cells died (Fig. 4.1). Using the OxICAT method, we monitored the redox status of 286 cysteine-containing peptides representing 224 proteins, including the majority of redox-sensitive peptides identified in our previous study (Chapter 3).

We found that in midlogarithmic phase, the majority of protein thiols were in their reduced state, showing less than 15% oxidation (Appendix, Table 4.A). This result agreed well with previous studies by us (Chapter 3) and others [263] and showed that protein oxidation is insignificant in exponentially growing cells in SCD media [263]. Most proteins maintained their low oxidation status for the first two days of incubation. Then, however, within a 24-hour period, more than 70% of identified peptides appeared to accumulate in their thiol-oxidized state. Nearly all of the peptides that followed this general oxidation trend showed an at least three-fold increase in thiol oxidation at day 3 of cultivation, and were fully oxidized by day 4 (Appendix, Table 4.A). Despite the high oxidation, almost 80% of yeast cells were still alive after four days of incubation (Fig. 4.1), suggesting that cysteine oxidation significantly precedes cell death. These results suggested a general collapse of the cellular redox balance during chronological aging. While cells in exponential or early stationary phase have a high antioxidant capacity, older yeast cells seem to lose the capacity to maintain their cellular redox homeostasis, resulting in increased cysteine oxidation.

4.3.4 Reduction in glucose concentration causes delayed collapse of the general redox balance

As described previously and shown in Fig. 4.1, the chronological life span of yeast cells was significantly extended by lowering the initial glucose concentration in the medium. To analyze if conditions that extend CLS also influence the thiol redox status in yeast, *S. cerevisiae* was cultivated in 0.5% glucose CR media and OxICAT was again applied to quantitatively assess the redox status of the proteins *in vivo*.

As before, the majority of peptides were found to be in a thiol-reduced state during midlogarithmic phase (Appendix, Table 4.B). Importantly, the oxidation status of peptides in CR media was very similar to the redox status of yeast cells cultivated in SCD media, demonstrating not only that the OxICAT technique is highly reproducible but confirming recent studies that showed comparably low protein oxidation in yeast cells cultivated in SCD or CR media [263]. ROS formation, which is induced during the diauxic shift from fermentation to respiration and thus occurs earlier in low-glucose CR media [3], seems to be offset by an increased antioxidant capacity of cells cultivated in CR media [365-368]. These observations again highlight the importance of the cellular antioxidant machinery, which maintain the redox balance during elevated ROS production in midlogarithmic growth.

Analysis of the redox status of proteins during CLS demonstrated that the collapse of the cellular redox status also occurs in CR media but with a considerable later onset. The collapse of the cellular redox status was about 48 hours delayed as compared to the collapse observed in yeast cells cultivated in SCD media and occurred between day 4 and 5 of cultivation (Appendix, Table 4.C), Noteworthy, oxidation affected mostly the same cysteine-containing peptides, indicating that this response is conserved. The delayed onset of cysteine oxidation appeared to correlate well with the extended life span of yeast cells, suggesting that maintenance of the cellular redox balance might indeed contribute to the life expanding benefits of reducing caloric intake.

Cultivation of yeast cells in SCD media for two days followed by a shift into water reflects an extreme caloric restriction measure, extending mean life span to 16 days (Fig. 4.1) and CLS for up to 25 days [342,343,352,353]. We monitored thiol oxidation under these conditions for 10 days following the shift and failed to observe any significant protein oxidation (Appendix, Table 4.C). Many proteins showed only a very minor increase in thiol oxidation between day 2 and day 3 of cultivation and then maintained this slightly increased oxidation status for at least 10 days in water. This result is in stark contrast to the collapse of the cellular redox homeostasis observed in cells continuously cultivated in SCD or CR media and is in excellent correlation with the extended life span of these cells. The slight increase of cysteine oxidation that was observed within 24 hours after the shift into water might reflect the beginning of the collapse of the redox balance that was shown for cells left in SCD media.

4.3.5 Oxidation of the active site cysteines in GapDH illustrates the redox collapse during CLS

Oxidation of the active site cysteine of glyceraldehyde-3-phosphate dehydrogenase (GapDH) serves as a typical representative of the group of proteins that fall victim to the general collapse of the cellular redox homeostasis during aging. The redox-sensitive cysteine residues Cys150 and Cys154 [369], which are located in the peptide aa 144-160 of yeast GapDH, were mainly reduced during the first two days of incubation in SCD media (Fig. 4.2). When yeast cells were cultivated in 2% glucose media, the two cysteine residues became heavily oxidized after day 3 of cultivation, whereas in 0.5% glucose media, the onset of oxidation was delayed by ~48 hours (Fig. 4.2) Yeast cells cultivated in SCD for two days and then switched into water did not show any significant increase in oxidation for GapDH's Cys150 or Cys154 for the time span that was monitored with OxICAT (Appendix, Table 4.C).

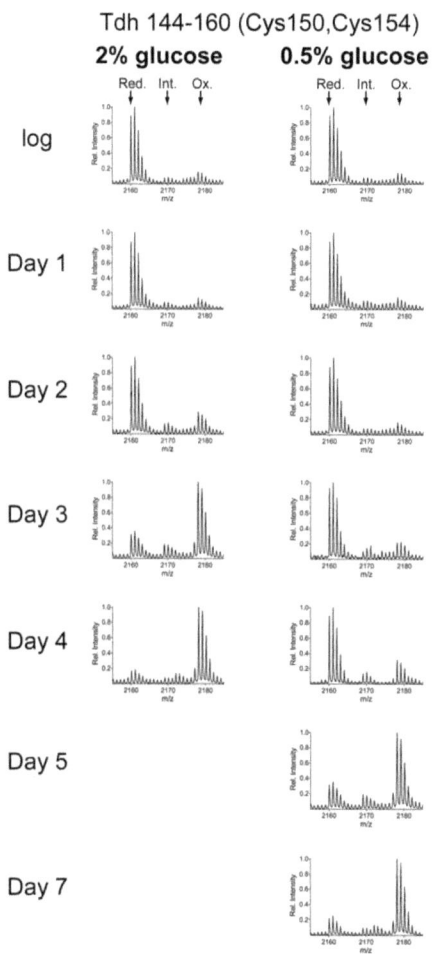

Figure 4.2. Oxidation of the redox-sensitive cysteines in GapDH illustrates the redox collapse during CLS. Shown are details of the mass spectrum of the glyceraldehydes-3-phosphate dehydrogenase (GapDH) isoenzyme Tdh2 or Tdh3 peptide 144-160 (peptide identical in both isoenzymes), which contains the redox-sensitive Cys150 and Cys154, during chronological yeast aging in 2% and 0.5% glucose media. The peptide elutes as an ICAT-triplet; the fully reduced peptide, which is labeled with two light ICAT molecules (calculated m/z = 2160.11), the fully oxidized peptide, which is labeled with two heavy ICAT molecules (m/z = 2178.18) and an oxidation intermediate, in which one cysteine is labeled with the light ICAT and one cysteine is labeled with the heavy ICAT (calculated m/z = 2169.15) (all values are given for Tdh3). GapDH exemplifies the onset of oxidation observed in the majority of identified proteins and the delay of oxidation induced by caloric restriction.

Oxidation of GapDH most likely results in disulfide bond formation between the redox-sensitive cysteines in GapDH and subsequent inactivation of the enzyme. GapDH inactivation is linked to the re-routing of the metabolic flux to the pentose phosphate pathway and increased NADPH production [107,108], which is used as important reducing equivalent in the cell. Thus, GapDH inactivation is considered to be a useful mechanism to increase the NAPDH output and enhance survival during oxidative stress [107,108]. However, based on the massive increase of oxidized proteins in aging yeast, it seems more likely that oxidation of GapDH is not used by cells as a regulatory tool but results from the collapse in cellular redox balance.

4.3.6 Clustering analysis reveals early targets of oxidation

The massive accumulation of oxidized proteins in yeast cells cultivated in SCD or CR media indicated a general collapse of the cellular redox balance. To determine whether all proteins follow the same pattern or whether oxidation of a few key proteins might contribute or even trigger the event, we conducted clustering analysis with all of the yeast peptides identified during chronological aging in SCD and CR media. Clustering is based on the algorithm *k-means with Euclidean distance*, implemented in *MultiExperimentViewer (MEV)*. The oxidation levels of the cysteine-containing peptides from chronological aging yeast cells in SCD and CR media served as input for the clustering analysis. For both media conditions, clustering analysis grouped the peptides into eight clusters (A-H), based on the trend of thiol oxidation during the course of the life span experiment (Fig. 4.3 and 4.4; all peptides in each cluster are listed in Appendix, Table 4.A and 4.B).

Clustering analysis confirmed that the majority of peptides followed a common thiol oxidation pattern and accumulated in the oxidized state between day 2/3 in SCD media or day 4/5 in CR media, respectively. These peptides were separated into three clusters, A-C (Fig. 4.3 and 4.4). While thiol oxidation in cluster A exceeded 50% at day 3 in SCD or day 5 in CR, oxidation of peptides in cluster B was less than 50% at that point and further increased at day 4 or 6, respectively. Peptides in cluster C revealed increased steady-state oxidation levels during logarithmic phase, and showed a further increase in oxidation at the time points listed for peptides in clusters A and B.

Peptides in cluster D and E have an onset of thiol-oxidation that preceded the oxidation of peptides in clusters A – C by either ~48 hours (cluster D) or ~24 hours (cluster E) (Fig. 4.3 and 4.4). For cells cultivated in SCD media, we found seven (Cluster D) and 18 peptides (Cluster E), respectively that displayed this early onset of oxidation. A very similar trend was observed in cells cultivated in CR, where nine and 16 peptides were found to cluster in groups D and E, respectively. 13 proteins that show early onset of thiol oxidation overlap between the two cultivation conditions (Table 4.1; Appendix, Table 4.C), suggesting that their oxidation might be a conserved event in CLS.

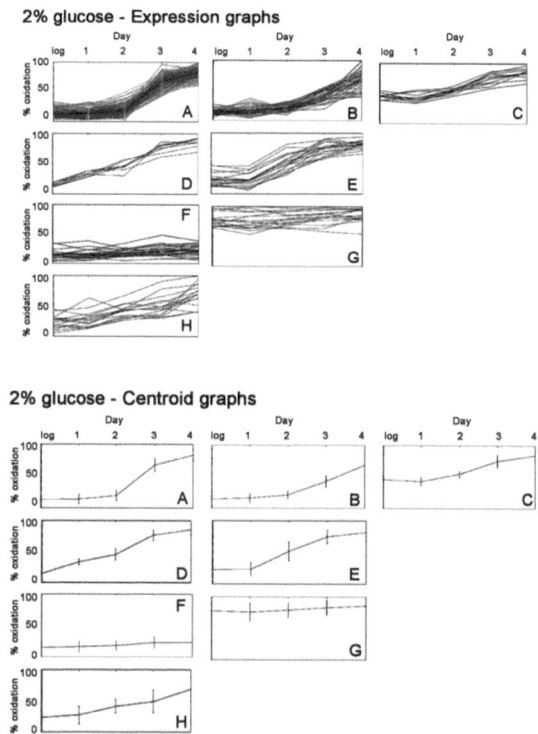

Figure 4.3. Clustering analysis of peptides identified during chronological aging in 2% glucose SCD media. *Peptides identified in chronological yeast aging (logarithmic phase – Day 4) in SCD media were clustered by the trend of their thiol oxidation during the course of the aging experiment. Eight individual clusters were found (labeled from A-H in the lower right corner of each graph), which are displayed as (upper panel) expression graphs and (lower panels) centroid graphs. Expression graphs display the oxidation pattern of each protein in this cluster across chronological aging (x-axis), with oxidation ranging from 0% to 100% (y-axis). In the centroid graph, individual lines are omitted and a single line, representing the mean oxidation status of all proteins within the cluster, is displayed. Error bars represent the standard deviation at a particular day within the cluster.*

*Cluster **A**: accumulate in thiol-oxidized state at day 3 & oxidation at day 3 > 50%; **B**: accumulate in thiol-oxidized state at day 3 & oxidation at day 3 < 50%; **C**: accumulate in thiol-oxidized state at day 3 & oxidation at log phase > 40%; **D**: onset of oxidation precedes oxidation of cluster A-C by 48 hours; **E**: onset of oxidation precedes oxidation of cluster A-C by 24 hours; **F**: remain reduced during the course of the experiment; **G**: remain oxidized; **H**: minor increase in oxidation over several days. The proteins represented by the peptides of each cluster are listed in Appendix, Table 4.A.*

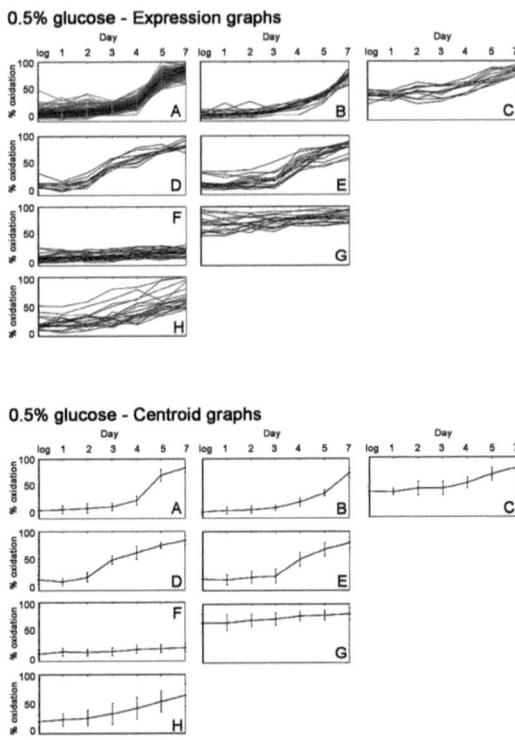

Figure 4.4. Clustering analysis of peptides identified during chronological aging in 0.5% glucose CR media. *Peptides identified in chronological yeast aging (logarithmic phase – Day 7) in CR media were clustered by the trend of their thiol oxidation during the course of the aging experiment. Eight individual clusters were found (labeled from A-H in the lower right corner of each graph), which are displayed as (upper panel) expression graphs and (lower panels) centroid graphs (see* Fig. *4.3 for more details).*

*Cluster **A**: accumulate in thiol-oxidized state at day 5 & oxidation at day 5 > 50%; **B**: accumulate in thiol-oxidized state at day 5 & oxidation at day 5 < 50%; **C**: accumulate in thiol-oxidized state at day 5 & oxidation at log phase > 40%; **D**: onset of oxidation precedes oxidation of cluster A-C by 48 hours; **E**: onset of oxidation precedes oxidation of cluster A-C by 24 hours; **F**: remain reduced during the course of the experiment; **G**: remain oxidized; **H**: minor increase in oxidation over several days. The proteins represented by the peptides of each cluster are listed in Appendix, Table 4.B.*

Chapter 4: Collapse of the cellular redox balance in aging yeast 135

Table 4.1. Yeast proteins preceding the general redox collapse in both SCD and CR media.
The % oxidation of the ICAT-labeled cysteine(s) is shown during yeast aging. Values that show a significant increase in thiol oxidation are printed in bold. GapDH and aconitase exemplify the general collapse of the redox balance during aging.

Gene, Peptide (Cys)	Protein	2% glucose						0.5% glucose						
		log	Day 1	Day 2	Day 3	Day 4	log	Day 1	Day 2	Day 3	Day 4	Day 5	Day 7
CCT4, 393-402 (399)	T-complex protein 1 subunit delta	12	24	**58**	87	86	12	14	13	11	**39**	72	82
CDC48, 104-119 (115)	Cell division control protein 48	12	29	**55**	75	93	13	15	14	28	**62**	68	90
IDP1, 393-403 (398)	Isocitrate dehydrogenase 1	18	19	**41**	61	85	16	18	32	25	**49**	81	80
KGD1, 982-996 (983)	2-oxoglutarate dehydrogenase E1	25	21	**49**	57	84	18	18	19	16	**57**	54	70
LYS2, 611-625 (614)	Aminoadipate-semialdehyde dehydrogenase	15	22	**49**	79	86	14	11	12	**48**	56	74	81
MES1, 353-366 (353)	Methionyl-tRNA synthetase	23	34	**64**	86	82	20	21	31	**61**	78	80	98
OLA1, 43-62 (43)	Uncharacterized GTP-binding protein	17	12	**46**	91	83	20	10	22	29	**53**	78	88
PRB1, 344-368 (361)	Cerevisin	18	23	**57**	91	94	12	14	26	23	**77**	80	88
PYC2, 211-222 (218)	Pyruvate carboxylase 2	11	**28**	**46**	81	93	13	11	22	**47**	66	77	81
RPL42B, 68-76 (74)	60S ribosomal protein L42	15	22	**49**	86	86	15	19	25	**49**	54	77	93
SES1, 411-433 (413;414)	Seryl-tRNA synthetase	22	23	**55**	66	79	20	19	17	**43**	65	71	81
TRR, 138-153 (142;145)	Thioredoxin reductase	33	33	**65**	77	82	34	39	42	44	**73**	80	86
UBC4, 103-129 (108)	Ubiquitin-conjugating enzyme E2 4	19	23	**48**	70	81	20	15	22	24	**49**	73	88
TDH, 144-160 (150;154)	Glyceraldehyde-3-phosphate dehydrogenase	26	22	28	**70**	81	24	18	21	26	28	**72**	87
ACO1, 376-392 (382)	Aconitase 1	7	4	12	**69**	84	7	11	11	20	28	**84**	88

4.3.7 Oxidation of thioredoxin reductase precedes general redox collapse

It is known that proteins differ strongly in their susceptibility to oxidative damage. For instance, redox-sensitive amino acids of bovine serum albumin were shown to be oxidized about twice as fast as those of glutamine synthase [370]. Clustering analysis of the proteins identified in this study revealed that the oxidation of only a few proteins substantially preceded the general collapse of the cellular redox homeostasis in chronologically aging yeast cells (Table 4.1). One of these early oxidation targets was the highly conserved thioredoxin reductase, Trr, a key component of the cellular redox-balancing machinery. The main role of cytosolic Trr1 and mitochondrial Trr2 is to regenerate thioredoxin (Trx), which maintains numerous cytosolic and mitochondrial proteins in their reduced, active state [371] (Fig. 4.5). Thioredoxin reductase accomplishes this by drawing electrons from the cellular NADPH pool.

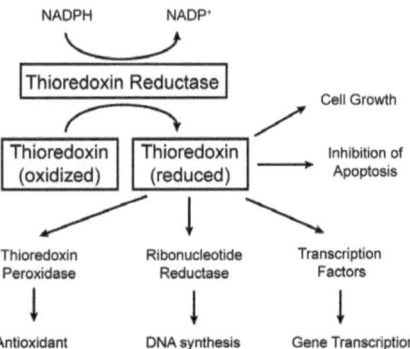

Figure 4.5. The thioredoxin system. *Thioredoxin reductase utilizes NADPH to catalyze the conversion of oxidized thioredoxin (Trx) into reduced Trx. Reduced Trx provides reducing equivalents to (i) Trx peroxidase, which breaks down H_2O_2 to water, (ii) ribonucleotide reductase, which reduces ribonucleotides to deoxyribonucleotides for DNA synthesis and (iii) transcription factors, which reveal increased binding to DNA and altered gene transcription when reduced. In addition, Trx promotes cell growth and inhibits apoptosis (Figure adapted from [372]).*

Because the identified OxICAT peptide, which contains the active site cysteines of Trr, is identical between the two homologues, we are unable to determine which of the two homologues becomes increasingly oxidized in chronologically aging yeast cells. However, Trr1 is present in yeast cells with ~290,000 molecules while mitochondrial Trr2 only has ~400 molecules [373], making it virtually impossible to detect and quantify Trr2 peptides by mass spectrometry. It is thus more likely that the majority of the peptide identified to be oxidized in this study was derived from cytosolic Trr1.

Thioredoxin reductase showed a basal level of ~35% oxidation in exponentially growing yeast cells. Such elevated oxidation can be explained by Trr's role in maintaining the redox homoeostasis of the cell by re-establishing the reduced state of thioredoxin (Fig. 4.5). During chronological aging, the onset of Trr oxidation preceded the onset of oxidation of the majority of cellular proteins (Fig. 4.6, compare with Fig. 4.2). In SCD media, the Trr peptide started to shift towards an oxidized population at day 2 of the aging yeast culture, about 24 hours before the major redox collapse was observed (Fig. 4.6). The same trend was observed for Trr in CR media, where the onset of Trr oxidation was at day 4, about 24 hours before the general collapse at day 5 (Fig. 4.6). These results suggest that in the first days of growth, Trr is capable of keeping the vast majority of proteins in their reduced and active state. However, increasing amounts of ROS in aging yeast cells might cause oxidation and subsequent loss of activity or even complete inactivation of Trr. Inactivation of Trr would result in the incapacity of Trr to reduce Trx, and likely explain the observed general collapse of the cellular redox balance and a shift towards an oxidative environment in the cell. Thus, early oxidation and inactivation of thioredoxin reductase may not only serve as a bellwether for an increase in oxidative damage but might be directly involved in determining the life span of yeast.

Noteworthy, we found that the early oxidation of thioredoxin reductase was partially reversible. When yeast cells were transferred into water after two days cultivation in SCD media, about 60% of the Trr molecules were found to be oxidized. However, one day of incubation in water resulted in a significant re-reduction of Trr (oxidation status: 43%) (Fig. 4.6; Appendix, Table 4.C). This oxidation status was maintained for the next 10 days in which we monitored yeast cells incubated in water. These results strongly suggest that this early intervention restores the activity of thioredoxin reductase, prevents the collapse of the cellular redox homeostasis and leads to a significantly extended life span.

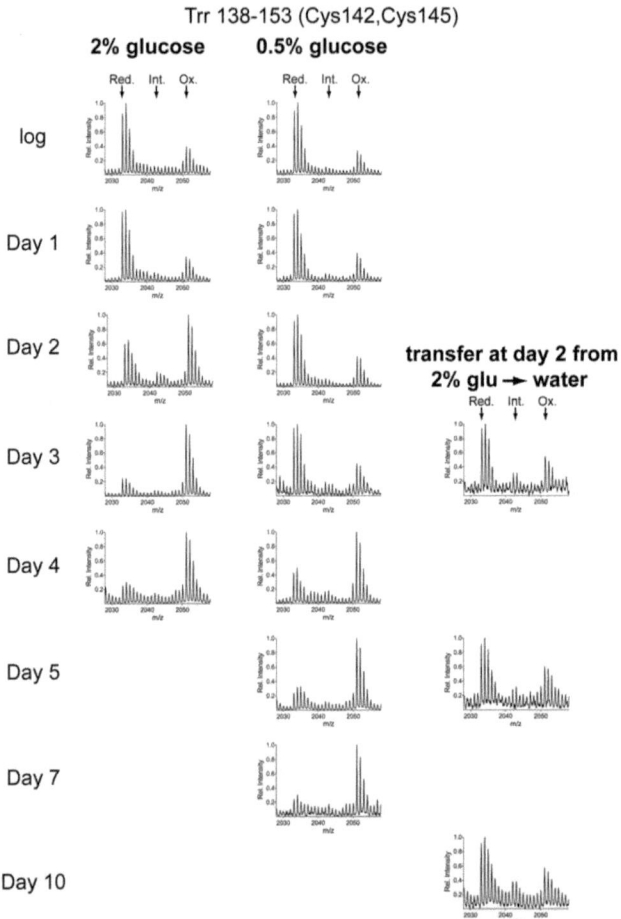

Figure 4.6. Thioredoxin reductase precedes general redox collapse. *Shown are details of the mass spectrum of the thioredoxin reductase (Trr) peptide 138-153, which contains the redox-sensitive Cys142 and Cys145, during chronological aging of yeast cells in i) 2% glucose SCD media, ii) 0.5% glucose CR media and iii) transferred into water after two days of incubation in 2% glucose SCD media. Like peptide aa 144-160 of GapDH (Fig. 4.2), the peptide elutes as an ICAT-triplet; the fully reduced peptide, which is labeled with two light ICAT molecules (calculated m/z = 2033.03), the fully oxidized peptide, which is labeled with two heavy ICAT molecules (m/z = 2051.09) and an oxidation intermediate, in which one cysteine is labeled with light ICAT and the other with heavy ICAT (calculated m/z = 2042.06) (all values are given for Trr1). In SCD and CR media, oxidation of the redox-sensitive cysteines in Trr precedes the general redox collapse, which is represented by GapDH in Fig. 4.2. For cells transferred into water, re-reduction of Trr and no significant increase in thiol oxidation over time was observed.*

According to our hypothesis, early oxidation of Trr might have important implications for the life span of chronologically aged yeast cells. Thus, if not compensated by the upregulation of other redox balancing systems, deletion of *trr* should cause an even earlier onset of oxidation and potentially a decrease in CLS of *S. cerevisiae*. Yeast cells lacking cytosolic *trr1* or mitochondrial *trr2* have been shown to suffer from increased oxidation but are viable [371]. Our group tested the viability of chronological aging Δ*trr1* and Δ*trr2* yeast single deletion strains in the same *EG103* strain background that we used for our previous life spans. Preliminary results showed that both *trr* deletions caused a decrease in the mean life span of yeast cells cultivated in SCD media, while no effect was observed in CR media (Sebastian Brandhorst, personal communication), suggesting that strains lacking thioredoxin reductase are no longer capable of maintaining their redox homeostasis. It still remains to be tested why the life span was unaffected under CR conditions. It is feasible that a transcriptional response, involving the expression of independent reducing systems and ROS scavengers, such as the GSH-glutaredoxin pathway, catalase and superoxide dismutase, might restore survival rates to wild type conditions.

The physiological events that trigger the early oxidation of thioredoxin reductase remain still to be elucidated as well. As mentioned previously, thioredoxin reductase draws its electrons from NADPH, making the redox status of the enzyme ultimately dependent on cellular NADPH/NADP$^+$ ratios [374]. It is known that expression of at least one enzyme involved in NADPH regeneration, the cytoplasmic NADP$^+$-dependent isocitrate dehydrogenase Idp2, is repressed by glucose and induced upon the diauxic shift [375]. This observation would suggest that yeast cells cultivated under conditions of caloric restriction have higher Idp2 levels and NADPH/NADP$^+$ ratios than yeast cells grown under standard conditions, and might explain why thioredoxin reductase oxidation is delayed in caloric restricted yeast cultures. Thus, in the future, we will determine NADPH/NADP$^+$ levels and analyze whether changes in cellular NADPH levels correlate with the oxidation and inactivation of thioredoxin reductase. It is interesting to note that Cys398 of mitochondrial Idp1 was identified as an early oxidation target in both SCD and CR media, suggesting its early inactivation. However, studies in *S. cerevisiae* suggested that it is mainly the activity of Idp2 and not of Idp1 that protects yeast cells against peroxide stress by supplying NADPH to thiol-dependent mechanisms that remove H_2O_2 [376]. Thus, oxidation and potential inactivation of Idp1 might have only a minor influence on the cellular NADPH level.

Future aims will also involve the generation of a Δ*trr1*Δ*trr2* double deletion to determine CLS and onset of redox collapse in this strain. Moreover, we will identify thioredoxin-dependent proteins, whose oxidation and, potentially, inactivation in the absence of a functional thioredoxin system might be directly involved in life span determination. These follow-up experiments will significantly enhance our understanding about the importance of maintaining cellular redox homeostasis during aging and specifically elucidate the role that thioredoxin reductase plays in this process.

4.3.8 Other early targets of oxidation in chronologically aging yeast

4.3.8.1 ATP-dependent molecular chaperone Cdc48

Protein clustering revealed several other important target proteins, whose oxidation precedes the general collapse of the cellular redox homeostasis in chronological aging yeast in both SCD and CR media (Table 4.1). One of these early targets of oxidation is Cdc48 (*i.e.*, p97, VCP) [377], an ATP-dependent molecular chaperone, which is involved in a number of essential cellular functions, including membrane fusion, protein degradation, transcriptional activation, cell cycle control and apoptosis (Fig. 4.7) [1]. Cdc48 harbors several cysteine residues, of which only three are conserved between yeast Cdc48 and human VCP [378]. Interestingly, one of these conserved cysteines is the highly oxidation sensitive Cys115 (Table 4.1), which has been found to be sensitive to peroxide-mediated oxidation by us and others (Chapter 3; [245]). So far, it is unclear how oxidation of Cys115, which resides in the N-terminal substrate-binding domain of Cdc48, affects the functionality of the chaperone and/or the chronological life span of yeast. It is interesting to note, however, that the ATPase activity of VCP and Cdc48 appears to be redox-regulated [379], but it is unclear whether oxidation of Cys115 plays a role in this process.

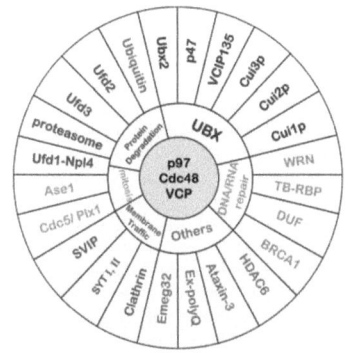

Figure 4.7. Cellular functions of Cdc48. *Classification of Cdc48 interacting proteins according to structural (UBX) or general functional criteria (protein degradation, mitosis, membrane traffic and DNA/RNA repair) (Figure adapted from [1]).*

4.3.8.2 Semialdehyde dehydrogenase Lys2

Interestingly, nine of the 35 NADPH-utilizing enzymes that exist in *S. cerevisiae* according to SGD were identified in our study. Oxidation of five of these enzymes (Erg9, Erg11, Fas2, Hom2, Ilv5) followed the general oxidation trend of cluster A, while cysteines in two of those NADPH-utilizing enzymes, Trr and L-aminoadipate-semialdehyde dehydrogenase Lys2, were oxidized early. Lys2 is involved in the biosynthesis of the essential amino acid L-lysine by reduction of α-aminoadipate to α-aminoadipate 6-semialdehyde, using NADPH as electron source [380]. Oxidation of Cys614 in Lys2 and potential inactivation of the enzyme might either occur as secondary effect due to limited amounts of available NADPH or might be a primary response to reduce NADPH consumption, making the reducing equivalent available to other cellular processes.

4.3.8.3 Pyruvate carboxylase Pyc2

Another early target of oxidation in both SCD and CR media was found to be Cys218 of pyruvate carboxylase Pyc2, a biotin-containing enzyme that catalyzes the carboxylation of pyruvate to oxalacetate. Pyc2 plays a role in various metabolic pathways including gluconeogenesis, lipogenesis, and porphyrinogenesis by replenishing the TCA cycle with oxalacetate [381]. In mammalian cells, changes in the amount of active Pyc2 are associated with the long-term regulation of metabolic flux from pyruvate to oxalacetate and with alterations in the concentrations of other key gluconeogenic enzymes [381]. It has been shown that the highly conserved Cys249 in Pyc1, a protein highly homologous to Pyc2, is involved in the binding of enzyme activators, including potassium ions and acetyl-CoA [382]. It is unclear whether the less conserved Cys218, which we find to be oxidized in our study, forms an intramolecular disulfide bond with Cy249 or a mixed disulfide with glutathione. If oxidation of Cys218 was to cause the inactivation of Pyc2, this would lead to a decrease in gluconeogenesis, which is needed during stationary growth to synthesize trehalose and glycogen. Since these storage metabolites are important for cell survival during stationary phase, a reduced synthesis might influence and limit the survival of cells during stationary growth.

4.3.8.4 T-complex protein 1 subunit delta Cct4

In both SCD and CR media, Cys399 in Cct4 was identified as early target of oxidation. Cct4 is the delta-subunit of the heterooligomeric chaperonin Cct, which is composed of 8 different subunits in *S. cerevisiae* and functionally related to the bacterial folding helper protein GroEL [383,384]. Cct has been shown to facilitate the folding of several proteins, including actin and tubulin both *in vitro* [385,386] and *in vivo* [387], suggesting that it plays a role in cytoskeleton organization and mitotic spindle formation [388,389]. It has also been shown that mutation of the highly conserved Cys450 to tyrosine causes a hereditary neuropathy in rats, raising the possibility that misfolding of Cct's target proteins causes this disease [390]. It remains to be seen whether oxidation of Cys399, which is less conserved within the Cct4 chaperonin family [391] might also cause defects in the folding capacity of the T-complex and results in increased amounts of misfolded proteins and/or misorganization of the actin/tubulin network.

4.3.8.5 Early oxidation of proteins involved in protein translation

Three of the 13 proteins that show early oxidation in aging yeast were involved in protein translation (Table 4.1), suggesting that this process might be one of the primary targets of oxidation. One of the affected proteins is methionyl-tRNA synthetase Mes1, which specifically attaches methionine residues to its cognate tRNAs [392]. Noteworthy, the oxidation sensitive Cys353 in Mes1 is part of a zinc-finger-like CX_2C-X_9-CX_2C motif. It has been shown for the homologous *E. coli* methionyl-tRNA synthetase MetRS, which is to 28.1% identical to yeast MetRS, that substitution of any of these cysteines in the zinc finger-like domain causes a strong decrease in the enzyme activity [393]. A similar decrease can therefore be anticipated for yeast Mes1, which likely down-regulates protein translation [394].

It is interesting to note that despite the potential down-regulation of this essential function so early in the life span, yeast cells apparently still have the capacities to survive for several days. This is in agreement with recent studies that have been shown that reduced protein translation actually increases longevity in *C. elegans* and the replicate life span of *S. cerevisiae* [395,396]. These observations led to the hypothesis that decreased translation results in the reduced production not only of correct proteins but, potentially even more importantly, of proteins with translation errors or irreversible oxidative damage [397,398]. Down-regulation of protein translation would allow the endogenous repair and degradation machinery to keep in pace with the removal of damaged proteins, leading to an improved protein homeostasis in older cells and an increase in life span [397,398].

4.3.9 Proteins that maintain their oxidation status during chronological yeast aging

Clustering analysis of the identified peptides revealed that about ~15% of the cysteine-containing yeast peptides maintain their low (cluster F, Fig. 4.3 and 4.4) or high oxidation status (cluster G, 4.3 and 4.4) during the course of the experiment, independent of the cultivation conditions. Cluster F contains 21 and 28 different proteins with predominantly reduced cysteines in SCD and CR media, respectively, which remain reduced during the course of the chronological life span experiment, suggesting that these cysteines are insensitive towards aging-induced oxidation. Although oxidation-insensitive proteins have been shown before in non-quantitative proteomics studies investigating yeast aging, these proteins were not identified making a comparison impossible at this point [263,281].

Analysis of the localization of these cysteines within the proteins might shed light into the questions why these cysteines are resistant towards aging-induced oxidation. One explanation might be inaccessibility of the cysteine residue, which could be buried deep in the protein structure. However, statistical analysis could not be conducted because the crystal structures of only four members of this cluster are currently available. Interestingly, two cysteine-containing peptides representing inositol-1-phosphate synthase (Ino1) and uracil phosphoribosyl-transferase (Fur1) did not display any significant oxidation in aging yeast cells although they were previously found to undergo significant peroxide-mediated thiol oxidation (Chapter 3; Appendix, Table 3.B). It is conceivable that the millimolar concentration of H_2O_2 that we used in our peroxide stress studies will not be reached in the vicinity of these two proteins during aging, leaving them in their reduced form.

4.3.10 Early oxidation targets — Cysteines in unfolded regions

We previously observed that reactive, H_2O_2-sensitive cysteine residues reside more often in unfolded amino acid regions than H_2O_2-insensitive cysteines (Chapter 3). To assess whether we see a similar trend among proteins that are early oxidation targets during CLS (cluster D and E), we investigated whether we detect any structural features that are common among cysteines of the 13 early oxidation targets in both SCD and CR conditions (Table 4.1) and less represented in peptides of cluster A (general redox collapse), H (remain oxidized) or F (remain reduced). We determined whether the amino acid stretches in which the identified cysteine is located are folded or unfolded using the FoldIndex prediction program.

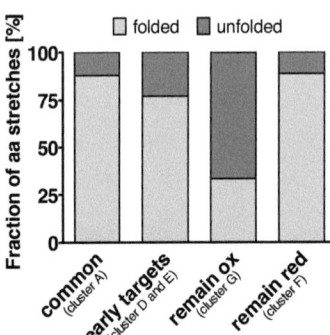

Figure 4.8. Fold propensity of different peptide clusters. *Fold propensity of all amino acid stretches in which a cysteine is located that either shows general redox collapse (cluster A; n = 124), early oxidation (Table 4.1; n = 13), remains oxidized (cluster H; n = 12) or remains reduced (cluster F; n = 18). Only the peptides that overlap in the respective clusters between SCD media and CR media were used for the determination of the fold propensity.*

The fold prediction revealed that cysteines that are early targets of oxidation or remain oxidized seem to be more often located in unfolded amino acid stretches than cysteines that follow the common redox collapse or remain reduced throughout the experiment (Fig. 4.8). Thus, cysteines that are located in unfolded amino acid stretches seem to be more prone to oxidation than cysteines in folded regions. Again, the low number of currently available crystal structures for proteins that are targets of early oxidation (n = 4) makes a proper statistical analysis of the accessibility of the cysteine within the protein (buried, bonded or on the surface) impossible.

4.3.11 Metabolic and redox metabolomic changes in aging yeast cells

4.3.11.1 Analysis of the glucose level in chronological aging yeast

As described previously and shown in Fig. 4.1, CLS can be significantly increased by lowering the glucose concentration in the medium. To assess glucose consumption in these cultures and to potentially correlate it to the life span of a chronological aging yeast population in 2% glucose SCD media or 0.5% glucose CR media, we determined the glucose levels during the first 24 hours of growth. In SCD media, glucose was exhausted within 14 hours of the initial inoculation (OD_{600} ~ 3.5) (Fig. 4.9). In contrast, in CR media, glucose was already depleted after 8 hours (OD_{600} of about 1). Importantly, the initial glucose concentration did not affect the growth rate of the culture (Fig. 4.9, compare closed circles and open squares). The observation that the initial glucose concentration in the media decisively influences CLS suggests that events occurring during the exponential growth phase or the diauxic shift have long lasting effects on the viability of a yeast population.

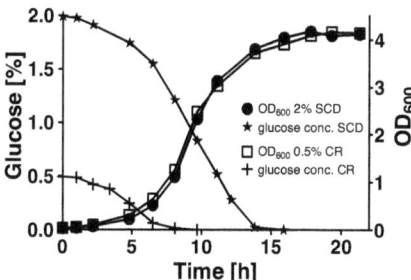

Figure 4.9. Optical density [OD_{600}] and glucose consumption [%] of a *S. cerevisiae* population cultivated in 2% glucose SCD media or 0.5% glucose CR media.

4.3.11.2 ATP level in chronological aging yeast

Despite limited supply of nutrients in stationary phase, it has been shown by oxygen consumption measurements that chronologically aging yeast cells are metabolically active in this phase [268,340]. To assess how intracellular ATP concentration change during CLS and to detect any potential correlation between the ATP levels and the oxidation status of proteins in yeast cells cultivated in SCD and CR media, we determined the intracellular ATP in aging yeast cells.

We found that in the first two days of cultivation, the absolute ATP levels in cells cultivated in 2% glucose SCD media were significantly higher compared to the ATP levels in cells cultivated in 0.5% glucose CR media (Fig. 4.10). These elevated ATP levels might be explained by the presence of a higher amount of metabolizable carbons under non-restricted conditions. No apparent difference was observed after this time point, suggesting that neither absolute ATP levels nor their dynamics are responsible for the delayed oxidation in glucose-restricted cells. Changes in ATP levels seem to follow the general protein oxidation pattern observed for both glucose conditions, as ATP levels in CR media show a similar delay of decrease compared to standard conditions as we have seen for protein oxidation. This time-delayed decline is most likely caused by oxidation of proteins involved in ATP production, such as the here identified Pet9, Atp3, Qcr2 or Qcr7 (Appendix, Table 4.A and 4.B) [399-401].

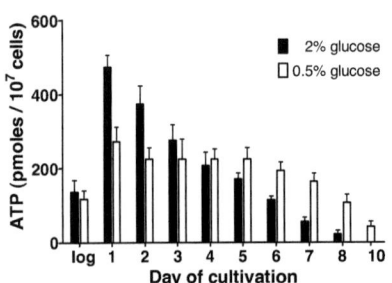

Figure 4.10. ATP levels in chronologically aging yeast cells cultivated in 2% glucose SCD media or 0.5% glucose CR media.

4.3.12 Analysis of intracellular GSH/GSSG and cysteine levels in chronological aging yeast

The small thiol-containing tripeptide glutathione, GSH and its oxidized counterpart GSSG, are highly abundant (~1-10 mM) and constitute the most important redox buffer of eukaryotic and prokaryotic cells [17]. During oxidative stress, glutathione maintains the redox state of cytosolic proteins by acting as an electron donor. In this process, GSH becomes oxidized to GSSG. It is then converted back to GSH by glutathione reductase (GR), which uses NADPH as ultimate electron donor [17]. Decreased levels of GSH are linked to increased ROS-mediated mitochondrial damage, apoptosis, and neurodegenerative diseases [402,403].

In order to understand how the redox potential of cells changes during the chronological aging process and to determine whether a correlation exists between the onset of protein oxidation and changes in the intracellular GSH and GSSG levels, we analyzed their concentrations in collaboration with the Banerjee lab here at the University of Michigan. Since the GSH/GSSG pool offers the greatest contribution to the redox environment, the redox potential E_h, which functions as an indicator of the redox environment of the cell, can be calculated by inserting the measured GSH and GSSG concentrations into the Nernst equation (*equation 1*). Under non-stress conditions, ~95-99% of the total glutathione pool is in the reduced form, constituting a redox potential of -220 mV to -240 mV in eukaryotic cells [242,404,405]. Under oxidative stress conditions, however, the redox potential drops below -200 mV [404].

During exponential growth, the intracellular concentrations of GSH and GSSG in *EG103* yeast cells were very similar in both SCD and CR media (Fig. 4.11A), leading to a calculated redox potential of ~ -225 mV (Fig. 4.11B). At day 1 in culture, however, the redox potential became significantly more oxidizing in SCD media as compared to CR media due to a strong increase in intracellular GSSG (Fig. 4.11A). After this time point, the GSSG concentration continuously increased with a comparable rate in SCD and CR media over time, while the GSH level remained similar up to day 4/5 in SCD and CR media, respectively, before it started to decrease. These changes in glutathione concentration resulted in redox potentials that reached comparable values with a delay of about 48 hour in CR media (Fig. 4.11B). This 48-hour difference is very similar to the observed delay in protein oxidation in CR media suggesting that changes in the GSH/GSSG pool either slightly precede or parallel protein oxidation. These observations are in agreement with previous studies that show a more oxidizing environment for cells in SCD media as compared to CR media after 72 hours of incubation [263].

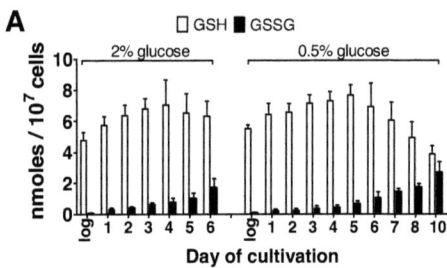

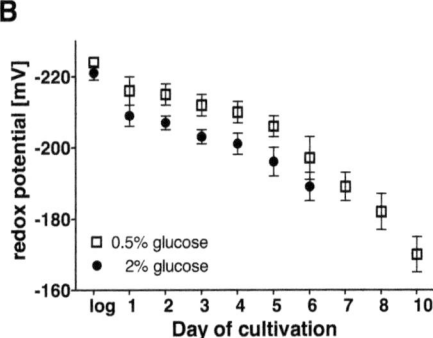

Figure 4.11. Intracellular GSH and GSSG concentrations and redox potentials in chronologically aging yeast cells in 2% glucose SCD media and 0.5% glucose CR media. *(A) Changes in the intracellular concentrations of GSH (white bars) and GSSG (black bars) during chronological yeast aging in SCD media and CR media. (B) Changes in the redox potentials during chronological yeast aging in SCD media (black dots) and CR media (white squares). Redox potentials were calculated by inserting the measured GSH and GSSG concentrations into the Nernst equation. Both GSH/GSSG concentrations and redox potentials were determined for up to 6 and 10 days in SCD media and CR media to match the time frame of the OxICAT experiments.*

The limiting factor in the biosynthesis of the tripeptide GSH is cysteine [406]. We measured the concentration of free intracellular cysteine and found a significant difference between yeast cells cultivated in SCD and CR media (Fig. 4.12) within the first 24 hours of cultivation. While the cysteine concentration dropped dramatically in SCD media, it increased between day 1 and day 2 in CR media (Fig. 4.12). These results show that yeast cells cultivated in CR media contain significantly more free cysteine than cells cultivated in SCD media. Cysteine is either synthesized in the cells [407] or taken up by cellular permeases from the media [408] and is used for protein synthesis, the formation of GSH and reversible cysteinylation of proteins. At this point, we are unable to make any assumptions, which of these processes might be affected in chronological aging yeast cells. It is, however, tempting to speculate that cultivation of yeast cells in SCD medium causes early protein oxidation, which leads to increased protein cysteinylation and a decrease in free levels of cysteines.

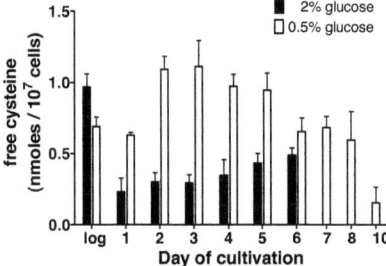

Figure 4.12. Cysteine concentration in chronologically aging yeast cells in 2% glucose SCD media and 0.5% glucose CR media. *Changes in the intracellular concentrations of free cysteine in 2% glucose SCD media (black bars) and 0.5% glucose CR media (white bars) during chronological yeast aging. Free cysteine concentration was measured for up to 6 and 10 days in SCD media and CR media to match the time frame of the OxICAT experiments.*

4.4 Conclusions

The present and upcoming enormous social and economic costs that arise due to an aging population have raised much interest in studying the mechanisms that underlie aging, a complex network associated with a progressive decline in cell integrity over time. Chronological aging of the budding yeast *S. cerevisiae* has been widely accepted as a model system to study aging because of its similarity to aging of post-mitotic mammalian cells [268]. Several studies have reported increasing levels of ROS and oxidized proteins as yeast cells age (7,9-12), supporting the "free radical theory of aging" which states that the progressive decline in the functional capacity of aging organisms is the result of the accumulation of oxidative damage caused by reactive oxygen species (ROS) [330]. However, none of these studies were able to quantitatively assess the extent or determine the precise onset of oxidative stress in aging yeast or any other organism [263,281].

In this study, we used a quantitative proteomic approach called OxICAT to precisely determine the extent and onset of protein thiol modifications in chronologically aging *S. cerevisiae*. Our data demonstrate that yeast cells suffer from a very abrupt loss in redox homeostasis, in which the large majority of cysteine-containing proteins become oxidized, and which precedes cell death by several days. While yeast cells in logarithmic phase have a large antioxidant capacity that allows them to rapidly recover from peroxide and superoxide stress treatment, chronologically aging yeast cells appear to lose this antioxidant capacity as they age. The onset of the redox collapse appears to correlate with the life span of the yeast cells as caloric restriction, which significantly extends CLS, delayed the redox collapse by about 48 hours. This suggests that maintenance of the redox balance might indeed contribute to the life expanding benefits of regulating the caloric intake of yeast.

Despite extensive studies [263,353,363,409], the exact mechanism(s) by which caloric restriction increases yeast life span has yet to be determined. The life span extension of yeast cells cultivated in 0.5% glucose media might be explained by the improved resistance of cells towards oxidative stress [365-368]. When glucose in the media is limited, yeast cells shift from fermentative to respiratory growth, the major source of ROS. This shift towards respiration, which occurs significantly earlier in CR media as compared to SCD media, results in enhanced oxygen consumption and is suggested to trigger increased ROS formation [3,351,409], which, in turn, is sufficient to induce antioxidant defenses, such as Sod1/Sod2, catalase, and GSH biosynthesis. These systems then help to offset the increased flux of ROS produced [365-368]. Thus, caloric restriction in yeast cells might promote better adaption to ROS early in life, which eventually could lead to an increased life span. Alternatively, it has been shown that stationary phase cells, initially cultivated under caloric restriction conditions, accumulate less iron and have show higher respiration rates and enhanced mitochondrial efficiency as compared to non-restricted conditions [267,281]. These parameters have been suggested to culminate in reduced leakage of electrons from the respiratory chain and, subsequently, a lower production of ROS [267,281]. These hypotheses are supported by increased resistance of midlogarithmic phase yeast cells cultivated in CR media towards H_2O_2 [281], reduced oxidative protein damage of stationary phase cells under CR conditions [281] and a higher level of cytosolic and mitochondrial ROS scavenging proteins in cells aging in CR media as compared to non-restricted cells [409]. It remains to be investigated whether it is the higher antioxidant capacity of CR-cultivated yeast cells and/or the lower ROS production that leads to the observed delay in thiol oxidation in stationary phase cells cultivated in CR media as compared to SCD media.

Not only the initial glucose concentration influences CLS and cysteine oxidation but also the media composition yeast cells are cultivated in after cell division has ceased. The shift of yeast cells that were cultivated in SCD media for two days into water resulted in extreme life span extension, indicating that yeast cells have the potential to modulate the length of their life span according to their environment. It also suggests that a decrease in metabolic rates, as observed under these cultivation conditions [268], extends life span, a conclusion that supports the free radical theory of aging. As both caloric restriction measures, either reducing the glucose concentration or shift into water, show extension of the chronological life span, it might be valid that these methods work in similar pathways and activate a common response to delay aging.

We observed that the free cysteine concentration in chronological aging yeast cells cultivated in SCD is significantly lower than in cells grown in CR media. Cysteine is the limiting factor in GSH biosynthesis [406]. Many studies have been reported increased S-glutathionylation, the reversible formation of mixed disulfides between protein thiols and GSH, in response to oxidative stress conditions [51,326]. A similar role has been shown for free cysteine, which is consumed during *S*-cysteinylation, the reversible modification of cysteine residues of proteins with free cysteine under conditions of oxidative stress

[143,410]. Both S-glutathionylation and *S*-cysteinylation have been shown to redox-modulate the activity of many proteins [143,209] and are suggested to fulfill a protective role against potential terminal oxidation of the cysteine residues during oxidative stress [51,143,326,410]. Thus, a higher concentration of free cysteines in CR media might lead to an increased rate of *S*-cysteinylation and a better protection against terminal protein thiol oxidation. As mechanisms that protect cells against oxidative stress have been considered to be important longevity-promoting factors [51,326], the free cysteine level might function as a type of redox regulation that co-determines onset of cysteine oxidation and life span of chronological aging yeast.

Although the higher level of free cysteine in cells cultivated in CR media did not translate into significantly higher GSH levels, analysis of the GSH/GSSG levels and redox potentials in chronological aging yeast cells cultivated in SCD and CR media revealed comparable redox potentials with a delay of about 48 hour in CR media (Fig. 4.11B). This 48-hour difference is very similar to the aforementioned delay in protein oxidation in CR media suggesting that changes in the GSH/GSSG pool either slightly precede or parallel protein oxidation. Together, these results indicate that the cellular environment changes into more oxidizing conditions earlier in cells cultivated in 2% glucose SCD media as compared to 0.5% glucose CR media, thereby setting the stage for wide-spread protein oxidation very early during the cultivation.

Clustering analysis revealed an overlapping subset of proteins in SCD and CR media, whose oxidative thiol modifications substantially precede the general redox collapse. These results suggest that the potential collapse of the cellular redox balance might be a programmed event with distinct proteins playing the predominant role during this process. Oxidation of these early target proteins, which most likely results in a loss of their activity or their inactivation, might contribute to or even cause the observed loss of redox homeostasis (*i.e.*, thioredoxin reductase) or cell death (*i.e.*, Cdc48). Oxidation of thioredoxin reductase (Trr) might lead to its inactivation and the incapacity of Trr to reduce thioredoxin (Trx), which would likely explain the massive accumulation of oxidized Trx target proteins inside the cell. Accumulation of large amounts of oxidized and potentially inactive proteins might be one of the underlying reasons for a shortened chronological life span of a yeast population.

We determined that the redox-sensitive cysteines of early oxidation targets are more often located in unfolded amino acids regions than cysteines of proteins that fall victim to the general redox collapse. It can be speculated that cysteines in unfolded regions are more likely to be exposed to ROS than cysteines in folded regions. It has been shown that not only protein thiols but also many other amino acid residues such as Met, His, Trp, and Tyr are suggested to react with oxidizing reactive species [323-325]. Thus, the oxidative load on individual amino acids might be higher in unfolded regions than in folded stretches, where neighboring amino acids might function as protective measure by participating in ROS scavenging, resulting in reduced ROS concentrations and lowered cysteine oxidation. Interestingly, the redox-sensitive cysteines of other known redox-regulated proteins such as the chaperone Hsp33 or the transcription factors RsrA, OxyR and Yap1 are all located in unfolded regions [29,43,50]. These results suggest that oxidation of proteins that precede the general redox collapse might be determined by the protein structure and less due to an unintentional ROS-induced process. Active targeting of critical redox-sensitive proteins by ROS might help cells save energy and reducing equivalents that are needed to counterattack oxidative stress.

Many of the oxidized proteins presented here have been shown before to contain H_2O_2 and/or superoxide-sensitive cysteines (Chapter 3; [51,245,263]. Short-term exposure of logarithmically growing cells to millimolar concentration of H_2O_2 or the superoxide-generating paraquat, however, caused less cysteine oxidation than age-induced oxidation of the same cysteine. This result most likely reflects the inactivation of redox balancing systems, such as Trr1, which rapidly restore redox homeostasis after a short bolus of oxidative stress in logarithmically growing cells and the presence of a more oxidizing redox potential.

Our results highlight the importance of a regulated redox-homeostasis during aging and showed that restricting the caloric intake results in delayed cysteine oxidation and increased survival rates of aging yeast. Identification of the affected proteins and determination of their degree of oxidative thiol modification during chronological aging provided us with valuable information about a subset of proteins whose thiol oxidation significantly preceded the global collapse of the cellular redox homeostasis. This information should allow us to explain some of the physiological alterations that are observed in aging cells and organisms. Future step will involve the investigation of the physiological events that lead to the early oxidation of selected *in vivo* protein targets and determine their precise role in the chronological aging process. These results will not only illustrate whether we developed an alternative approach to discover proteins involved in life span determination but will provide answers to the very fundamental question in aging research, how changes in the cellular redox status affect life span. They will form the foundation for future studies, in which we will test the hypothesis that the observed collapse of yeast's redox homeostasis might actually be a programmed event.

4.5 Appendix Chapter 4

Table 4.A. Thiol oxidation status of yeast proteins during chronological aging in 2% glucose SCD media. OxICAT was applied to chronological aging yeast cells cultivated in SCD media. Identified peptides are grouped according to their cluster (A-H). Gene designation, protein name, and Swiss-Prot accession numbers are based on annotations in SGD. "% oxidation" denotes the average oxidation of the redox-sensitive cysteine (second column, in brackets), obtained from three independent experiments ± standard deviation. *: peptide identical in both isoenzymes; a,b,c: at this mass, two different peptides were identified.

Cluster	Swiss-Prot	Gene, Peptide (Cys)	Protein name	log phase	% oxidation at...			
					Day 1	Day 2	Day 3	Day 4
A	*accumulate in thiol-oxidized state at day 3 & oxidation at day 3 > 50%*							
	P19414	ACO1, 376-392 (382)	Aconitase 1	7 ± 2	4 ± 3	12 ± 6	69 ± 7	84 ± 5
	P38009	ADE17, 285-304 (300)	Bifunctional purine biosynthesis protein	8 ± 4	10 ± 2	9 ± 6	79 ± 5	75 ± 5
	Q04894	ADH6, 141-174 (163)	NADP-dependent alcohol dehydrogenase 6	24 ± 3	23 ± 4	31 ± 12	50 ± 9	66 ± 6
	P47143	ADO1, 121-130 (121)	Adenosine kinase	11 ± 4	14 ± 6	12 ± 3	88 ± 4	92 ± 5
	P38013	AHP1, 114-124 (120)	Peroxiredoxin type-2	6 ± 2	5 ± 3	15 ± 11	73 ± 4	89 ± 5
	P38013	AHP1, 49-79 (62)	Peroxiredoxin type-2	12 ± 4	17 ± 4	18 ± 9	60 ± 7	80 ± 2
	P54115	ALD6, 124-134 (132)	Aldehyde dehydrogenase (Mg^{2+} dep.)	15 ± 4	14 ± 3	18 ± 6	66 ± 8	89 ± 1
	P54115	ALD6, 290-311 (306)	Aldehyde dehydrogenase (Mg^{2+} dep.)	15 ± 4	15 ± 3	29 ± 9	51 ± 7	67 ± 7
	P54783	ALO1, 359-374 (359)	D-arabinono-1,4-lactone oxidase	22 ± 4	33 ± 6	2 ± 2	81 ± 10	93 ± 5
	P16550	APA1, 127-139 (129)	AP4A phophorylase	7 ± 5	6 ± 2	6 ± 2	53 ± 5	87 ± 12
	P32449	ARO4, 240-249 (244)	Phospho-2-dehydro-3-deoxyheptonate aldolase	13 ± 6	17 ± 6	23 ± 7	56 ± 4	83 ± 10
	P49089	ASN1, 421-440 (432)	Asparagine synthetase 1	10 ± 4	14 ± 1	7 ± 2	67 ± 7	76 ± 4
	P49089/P49090	ASN1/2, 396-406 (404)*	Asparagine synthetase 1/2	7 ± 2	9 ± 1	15 ± 6	78 ± 7	83 ± 16
	P38891	BAT1, 125-141 (126)	Branched-chain-amino-acid aminotransferase	11 ± 5	9 ± 3	15 ± 8	72 ± 2	87 ± 6
	P38891	BAT1, 220-237 (227)	Branched-chain-amino-acid aminotransferase	8 ± 4	7 ± 2	22 ± 11	71 ± 15	90 ± 3
	P29311	BMH1, 77-88 (85)	14-3-3 protein	16 ± 4	17 ± 3	25 ± 11	84 ± 8	92 ± 3
	P39079	CCT6, 515-535 (530)	T-complex protein 1 subunit zeta	5 ± 1	4 ± 2	13 ± 6	63 ± 12	82 ± 14
	P39079	CCT6, 79-103 (96)	T-complex protein 1 subunit zeta	10 ± 4	12 ± 3	13 ± 7	62 ± 5	80 ± 8
	P00549	CDC19, 287-309 (296)	Pyruvate kinase 1	12 ± 3	15 ± 3	22 ± 8	73 ± 5	88 ± 11
	P00549	CDC19, 370-394 (371)	Pyruvate kinase 1	18 ± 5	22 ± 2	25 ± 11	84 ± 9	87 ± 9
	P25379	CHA1, 175-195 (178)	L-serine/threonine dehydratase	11 ± 3	11 ± 4	15 ± 6	53 ± 6	75 ± 8
	P03965	CPA2, 159-182 (170)	Carbamoyl-phosphate synthase large chain	10 ± 5	7 ± 4	12 ± 2	55 ± 7	79 ± 5
	P03965	CPA2, 673-694 (678)	Carbamoyl-phosphate synthase large chain	18 ± 2	21 ± 4	20 ± 9	81 ± 10	83 ± 3
	P53691	CPR6, 150-165 (156)	Peptidyl-prolyl cis-trans isomerase CPR6	7 ± 2	3 ± 1	6 ± 3	72 ± 6	93 ± 6
	P27614	CPS1, 361-370 (368)	Carboxypeptidase S	33 ± 3	22 ± 6	44 ± 3	77 ± 1	90 ± 6
	P38845	CRP1, 105-118 (113)	Uncharacterized	21 ± 3	16 ± 1	33 ± 13	67 ± 13	79 ± 5
	P24783	DBP2, 215-226 (219)	ATP-dependent RNA helicase DBP2	15 ± 2	22 ± 4	30 ± 3	74 ± 7	73 ± 13
	P06634	DED1, 256-270 (257)	ATP-dependent RNA helicase DED1	11 ± 5	11 ± 5	18 ± 4	83 ± 8	81 ± 5

UniProt	Gene (positions)	Description					
P04802	DPS1, 250-259 (255)	Aspartyl-tRNA synthetase	28 ± 4	22 ± 5	31 ± 7	76 ± 6	86 ± 12
Q06053	DUS3, 210-225 (224)	tRNA-dihydrouridine synthase 3	13 ± 4	19 ± 4	18 ± 3	74 ± 13	98 ± 3
P32324	EFT2, 353-370 (366)[a]	Elongation factor 2	12 ± 4	6 ± 2	13 ± 5	78 ± 6	88 ± 8
P32324	EFT2, 121-144 (136)	Elongation factor 2	11 ± 7	10 ± 1	26 ± 4	70 ± 10	86 ± 5
P32324	EFT2, 724-749 (735)	Elongation factor 2	8 ± 4	6 ± 4	15 ± 5	77 ± 12	85 ± 1
P00925	ENO2, 244-255 (248)	Enolase 2	5 ± 1	3 ± 2	10 ± 4	51 ± 6	87 ± 7
P38333	ENP1, 320-333 (330)	Essential nuclear protein 1	11 ± 3	4 ± 1	12 ± 5	67 ± 6	96 ± 3
P41338	ERG10, 343-362 (358)[b]	Acetyl-CoA acetyltransferase	12 ± 3	12 ± 2	19 ± 14	56 ± 11	82 ± 17
P10614	ERG11, 137-146 (142)	Cytochrome P450	15 ± 6	13 ± 3	8 ± 3	62 ± 8	81 ± 4
P29704	ERG9, 340-350 (341)	Squalene synthetase	9 ± 2	12 ± 3	24 ± 5	72 ± 7	62 ± 19
P19097	FAS2, 900-919 (917)	Fatty acid synthase subunit alpha	22 ± 3	16 ± 7	30 ± 17	62 ± 13	86 ± 5
P14540	FBA1, 209-230 (226)	Fructose-bisphosphate aldolase	12 ± 3	8 ± 3	22 ± 9	64 ± 10	79 ± 6
P14540	FBA1, 250-264 (254)	Fructose-bisphosphate aldolase	13 ± 3	16 ± 5	22 ± 9	63 ± 8	76 ± 10
P32614	FRDS1, 446-459 (454)	Probable fumarate reductase	6 ± 1	11 ± 4	25 ± 9	60 ± 2	90 ± 4
P15624	FRS1, 121-138 (137)	Phenylalanyl-tRNA synthetase beta chain	12 ± 5	15 ± 9	17 ± 4	74 ± 5	86 ± 4
P09032	GCD1, 163-188 (172)	Translation initiation factor elF-2B subunit gamma	13 ± 2	18 ± 3	16 ± 6	85 ± 4	79 ± 7
P33892	GCN1, 1070-1078 (1077)	Translational activator GCN1	22 ± 6	26 ± 9	33 ± 14	58 ± 5	71 ± 6
P39958	GDI1, 217-226 (221)	Rab GDP-dissociation inhibitor	14 ± 1	22 ± 2	20 ± 9	65 ± 4	89 ± 5
Q12154	GET3, 314-322 (317)	ATPase GET3	15 ± 4	15 ± 8	20 ± 7	50 ± 8	95 ± 2
P25370	GFD2, 88-103 (90)	Good for DBP5 activity protein 2	16 ± 7	13 ± 3	22 ± 5	58 ± 5	85 ± 6
P32598	GLC7, 20-40 (38)	Serine/threonine-protein phosphatase	9 ± 4	9 ± 1	24 ± 13	77 ± 8	74 ± 9
P38720	GND1, 287-295 (287)	6-phosphogluconate dehydrogenase	15 ± 7	9 ± 3	15 ± 2	54 ± 5	71 ± 5
P41277/P40106	GPP1/GPP2, 210-222 (211;211)*	(DL)-glycerol-3-phosphatase 1/2	27 ± 7	22 ± 1	38 ± 12	59 ± 14	69 ± 11
P16622	HEM15, 118-136 (123)	Ferrochelatase	27 ± 2	28 ± 7	39 ± 7	61 ± 6	77 ± 10
P13663	HOM2, 235-249 (247)	Aspartate-semialdehyde dehydrogenase	13 ± 3	11 ± 4	20 ± 9	68 ± 4	86 ± 6
P13663	HOM2, 250-263 (259)	Aspartate-semialdehyde dehydrogenase	20 ± 2	15 ± 2	19 ± 3	64 ± 8	76 ± 5
P10869	HOM3, 38-56 (53)	Aspartokinase	5 ± 1	6 ± 3	18 ± 4	67 ± 1	80 ± 7
P21954	IDP1, 360-371 (370)	Isocitrate dehydrogenase 1	10 ± 5	3 ± 2	9 ± 5	69 ± 12	84 ± 8
P21954	IDP1, 89-98 (89)	Isocitrate dehydrogenase 1	13 ± 3	9 ± 3	32 ± 4	65 ± 13	76 ± 5
P09436	ILS1, 99-120 (119)	Isoleucyl-tRNA synthetase	25 ± 5	22 ± 8	21 ± 4	95 ± 2	89 ± 4
P39522	ILV3, 157-180 (176)	Dihydroxy-acid dehydratase	14 ± 4	22 ± 3	27 ± 7	61 ± 5	73 ± 11
P06168	ILV5, 307-321 (316)	Ketol-acid reductoisomerase	15 ± 1	12 ± 3	17 ± 9	51 ± 5	82 ± 5
P25605	ILV6, 102-120 (111)	Acetolactate synthase small subunit	8 ± 2	17 ± 4	18 ± 8	65 ± 8	81 ± 9
P25605	ILV6, 261-272 (270)	Acetolactate synthase small subunit	9 ± 4	8 ± 3	17 ± 9	70 ± 9	84 ± 9
P25605	ILV6, 76-93 (82)	Acetolactate synthase small subunit	7 ± 3	5 ± 2	18 ± 11	69 ± 21	98 ± 1
P40069	KAP123, 977-987 (984)	Importin subunit beta-4	14 ± 3	17 ± 6	27 ± 10	67 ± 6	82 ± 11
P27809	KRE2, 123-130 (124)	Glycolipid 2-alpha-mannosyltransferase	33 ± 3	26 ± 5	49 ± 8	80 ± 9	97 ± 2
P15180	KRS1, 429-446 (438)	Lysyl-tRNA synthetase	9 ± 4	4 ± 2	18 ± 7	63 ± 11	78 ± 7
P15180	KRS1, 430-446 (438)	Lysyl-tRNA synthetase	10 ± 5	12 ± 3	22 ± 13	75 ± 4	86 ± 4
P15180	KRS1, 481-496 (486)	Lysyl-tRNA synthetase	13 ± 2	8 ± 3	16 ± 8	58 ± 13	68 ± 4
P07264	LEU1, 230-242 (233)	3-isopropylmalate dehydratase	15 ± 4	22 ± 3	29 ± 8	67 ± 4	80 ± 1
P07264	LEU1, 60-77 (63)	3-isopropylmalate dehydratase	22 ± 4	20 ± 4	24 ± 5	62 ± 7	85 ± 4
P06208	LEU4, 345-353 (345)	2-isopropylmalate synthase	10 ± 3	14 ± 5	18 ± 7	67 ± 4	79 ± 8
P09624	LPD1, 45-58 (54)	Dihydrolipoyl dehydrogenase	12 ± 2	17 ± 7	24 ± 4	72 ± 5	80 ± 7

ID	Gene	Description					
P53312	LSC2, 365-377 (365)	Succinyl-CoA ligase subunit beta	12 ± 4	10 ± 5	13 ± 6	68 ± 3	81 ± 12
P36013	MAE1, 535-555 (541)	Malic enzyme	10 ± 3	9 ± 3	19 ± 6	67 ± 4	86 ± 2
P53091	MCM6, 813-832 (815)	DNA replication licensing factor MCM6	20 ± 5	15 ± 3	16 ± 3	50 ± 6	75 ± 5
P05694	MET6, 563-573 (571)	Methionine synthase	14 ± 3	9 ± 3	26 ± 10	64 ± 13	89 ± 6
P53875	MRPL19, 36-46 (40)[c]	54S ribosomal protein L19	7 ± 3	7 ± 3	6 ± 4	60 ± 15	87 ± 4
P53081	NIF3, 105-123 (114)	NGG1-interacting factor 3	10 ± 3	7 ± 2	16 ± 5	68 ± 20	64 ± 4
P15646	NOP1, 273-288 (275)	rRNA 2'-O-methyltransferase fibrillarin	9 ± 3	15 ± 3	12 ± 5	73 ± 4	92 ± 7
P15646	NOP1, 312-319 (314)	rRNA 2'-O-methyltransferase fibrillarin	9 ± 7	9 ± 4	9 ± 4	69 ± 13	83 ± 14
P27476	NSR1, 233-254 (241)	Nuclear localization sequence-binding protein	14 ± 6	31 ± 9	17 ± 7	73 ± 12	77 ± 11
Q02630	NUP116, 1030-1042 (1031)	Nucleoporin NUP116/NSP116	9 ± 6	17 ± 5	21 ± 9	52 ± 7	70 ± 7
P38219	OLA1, 126-143 (126)	Uncharacterized GTP-binding protein	5 ± 3	6 ± 3	13 ± 6	81 ± 12	76 ± 14
P38219	OLA1, 72-81 (76)	Uncharacterized GTP-binding protein	9 ± 3	9 ± 5	13 ± 4	60 ± 18	83 ± 9
P06169	PDC1, 213-224 (221,222)	Pyruvate decarboxylase isozyme 1	22 ± 4	18 ± 6	13 ± 5	58 ± 3	61 ± 6
P16467	PDC5, 128-154 (152)	Pyruvate decarboxylase isozyme 2	11 ± 8	15 ± 6	12 ± 3	77 ± 10	78 ± 4
P18239	PET9, 66-76 (73)[c]	ADP,ATP carrier protein 2	7 ± 3	6 ± 5	10 ± 6	51 ± 19	80 ± 13
P18239	PET9, 265-273 (271)	ADP,ATP carrier protein 2	16 ± 2	14 ± 2	18 ± 6	75 ± 12	89 ± 6
P18239	PET9, 67-76 (73)	ADP,ATP carrier protein 2	13 ± 2	15 ± 6	12 ± 3	77 ± 6	71 ± 4
P07274	PFY1, 77-91 (89)	Profilin	7 ± 4	5 ± 2	10 ± 2	72 ± 15	92 ± 3
P00560	PGK1, 83-108 (98)	Phosphoglycerate kinase	9 ± 3	5 ± 3	8 ± 4	62 ± 6	86 ± 9
P37012	PGM2, 372-389 (376)[b]	Phosphoglucomutase-2	12 ± 4	9 ± 2	13 ± 7	87 ± 6	82 ± 13
P05030/P19657	PMA1/2, 392-414 (376,405)*	Plasma membrane ATPase 1/2	13 ± 3	12 ± 2	18 ± 5	72 ± 8	62 ± 7
P05030/P19657	PMA1/2, 415-443 (409,438)*	Plasma membrane ATPase 1/2	12 ± 7	16 ± 6	16 ± 4	68 ± 8	85 ± 4
P05030/P19657	PMA1/2, 487-503 (472;501)*	Plasma membrane ATPase 1/2	18 ± 5	17 ± 5	23 ± 11	69 ± 15	97 ± 1
P04840	POR1, 125-132 (130)	Mitochondrial outer membrane protein porin 1	8 ± 4	11 ± 3	14 ± 3	77 ± 2	86 ± 4
P53043	PPT1, 244-262 (247)	Serine/threonine-protein phosphatase T	12 ± 8	15 ± 4	27 ± 3	52 ± 8	70 ± 8
P21242	PRE10, 73-86 (76)	Proteasome component C1	12 ± 4	13 ± 7	17 ± 7	69 ± 5	73 ± 4
P41940	PSA1, 100-125 (113)	Mannose-1-phosphate guanyltransferase	12 ± 3	4 ± 2	11 ± 5	56 ± 12	77 ± 4
P07257	QCR2, 129-152 (137)	Cytochrome b-c1 complex subunit 2	23 ± 6	32 ± 7	35 ± 12	84 ± 7	90 ± 7
P32628	RAD23, 364-371 (365)	UV excision repair protein RAD23	23 ± 9	30 ± 10	26 ± 5	50 ± 11	82 ± 10
P11745	RNA1, 169-182 (180)	Ran GTPase-activating protein 1	14 ± 2	16 ± 4	27 ± 8	78 ± 9	90 ± 6
P49723	RNR4, 221-235 (228)	Ribonucleoside-diphosphate reductase small chain 2	12 ± 3	19 ± 2	19 ± 10	67 ± 4	81 ± 2
P41805	RPL10, 41-69 (49)	60S ribosomal protein L10	15 ± 5	15 ± 6	31 ± 12	73 ± 12	89 ± 4
P41805	RPL10, 70-82 (71)	60S ribosomal protein L10	21 ± 4	16 ± 5	23 ± 2	71 ± 9	86 ± 9
P17079	RPL12B, 131-142 (141)	60S ribosomal protein L12	8 ± 3	7 ± 3	15 ± 6	57 ± 2	75 ± 9
Q12672	RPL21B, 101-108 (101)	60S ribosomal protein L21-B	9 ± 3	9 ± 3	21 ± 10	62 ± 13	88 ± 3
P14796	RPL40A, 39-48 (39)	60S ribosomal protein L40	18 ± 3	21 ± 7	37 ± 7	73 ± 13	90 ± 3
P46654	RPS0B, 135-165 (162)	40S ribosomal protein S0-B	18 ± 8	17 ± 8	23 ± 7	70 ± 3	87 ± 8
P26781	RPS11B, 117-133 (128)	40S ribosomal protein S11	18 ± 2	15 ± 6	24 ± 9	78 ± 8	77 ± 9
P39516	RPS14B, 72-83 (72)	40S ribosomal protein S14-B	12 ± 6	14 ± 4	19 ± 4	73 ± 11	88 ± 4
P14127	RPS17B, 34-45 (35)	40S ribosomal protein S17-B	8 ± 3	5 ± 2	6 ± 3	61 ± 4	89 ± 1
P23248	RPS1B, 65-83 (69)	40S ribosomal protein S1-B	8 ± 3	5 ± 2	6 ± 2	60 ± 6	76 ± 6
Q3E754	RPS21B, 16-22 (17)	40S ribosomal protein S21-B	20 ± 6	15 ± 2	16 ± 3	56 ± 10	65 ± 3
P05759	RPS37, 68-73 (68)	40S ribosomal protein S31	24 ± 7	18 ± 2	25 ± 9	55 ± 13	76 ± 12
P05754	RPS8B, 165-178 (168)	40S ribosomal protein S8	9 ± 4	15 ± 11	18 ± 6	68 ± 9	84 ± 8

Accession	Gene, positions	Description					
Q12464	RVB2, 221-232 (224)	RuvB-like protein 2	13 ± 6	19 ± 5	27 ± 3	63 ± 7	74 ± 10
P39954	SAH1, 192-199 (198)	Adenosylhomocysteinase	16 ± 4	33 ± 6	28 ± 3	50 ± 4	84 ± 2
P39954	SAH1, 218-236 (231)	Adenosylhomocysteinase	12 ± 2	27 ± 9	18 ± 10	86 ± 6	86 ± 13
P21243	SCL1, 63-77 (74)	Proteasome component C7-alpha	4 ± 1	20 ± 3	14 ± 4	71 ± 9	88 ± 7
P47052	SDH1b, 613-632 (624)	Succinate dehydrogenase flavoprotein subunit 2	21 ± 4	15 ± 5	15 ± 7	72 ± 7	83 ± 6
P33330	SER1, 182-202 (182)	Phosphoserine aminotransferase	11 ± 3	9 ± 2	23 ± 9	55 ± 5	86 ± 8
P07284	SES1, 363-379 (370;373)	Seryl-tRNA synthetase	26 ± 4	30 ± 8	44 ± 4	69 ± 7	70 ± 6
P37292	SHM1, 89-106 (103)	Serine hydroxymethyltransferase	14 ± 2	12 ± 0	19 ± 10	50 ± 4	83 ± 7
P37291	SHM2, 188-198 (197)	Serine hydroxymethyltransferase	23 ± 7	15 ± 3	28 ± 7	62 ± 19	87 ± 9
P34223	SHP1, 78-103 (88)	UBX domain-containing protein 1	13 ± 3	9 ± 1	18 ± 5	51 ± 10	79 ± 7
P10592	SSA2, 14-34 (15)	Heat shock protein SSA2	17 ± 5	17 ± 3	13 ± 3	73 ± 5	99 ± 0
P10592	SSA2, 304-313 (308)	Heat shock protein SSA2	7 ± 1	11 ± 2	5 ± 3	81 ± 11	87 ± 7
P11484	SSB1, 431-451 (435)	Heat shock protein SSB1	14 ± 3	15 ± 6	22 ± 8	71 ± 8	86 ± 3
P40150	SSB2, 452-466 (454)	Heat shock protein SSB2	11 ± 3	7 ± 2	12 ± 3	73 ± 16	81 ± 13
P32590	SSE2, 375-388 (380)	Heat shock protein homolog SSE2	9 ± 4	9 ± 3	12 ± 7	58 ± 5	76 ± 8
Q07478	SUB2, 359-368 (360)	ATP-dependent RNA helicase SUB2	10 ± 5	19 ± 3	16 ± 6	54 ± 20	94 ± 3
P00359/P00358	TDH2/3, 144-160 (150;154)*	Glyceraldehyde-3-phosphate dehydrogenase 2/3	26 ± 3	22 ± 2	28 ± 7	70 ± 8	81 ± 7
P02994	TEF1, 21-37 (31)	Elongation factor 1-alpha	18 ± 3	14 ± 3	23 ± 11	62 ± 6	81 ± 1
P02994	TEF1, 320-328 (324)	Elongation factor 1-alpha	11 ± 4	8 ± 4	22 ± 8	85 ± 7	80 ± 4
P16521	TEF3, 665-696 (691)	Elongation factor 3A	11 ± 2	17 ± 8	12 ± 5	50 ± 6	68 ± 11
P16521	TEF3, 708-732 (731)	Elongation factor 3A	9 ± 4	16 ± 6	21 ± 7	80 ± 8	87 ± 4
P17255	TFP1, 356-367 (358)	V-type proton ATPase catalytic subunit A	9 ± 3	13 ± 5	22 ± 7	74 ± 7	88 ± 10
P17423	THR1, 216-231 (216)	Homoserine kinase	8 ± 3	6 ± 1	11 ± 3	60 ± 11	98 ± 1
P16120	THR4, 416-432 (428)	Threonine synthase	9 ± 3	9 ± 3	13 ± 4	53 ± 10	82 ± 8
P04801	THS1, 264-278 (268)	Threonyl-tRNA synthetase	26 ± 4	30 ± 11	32 ± 11	52 ± 9	60 ± 10
P38912	TIF11, 83-101 (89)	Eukaryotic translation initiation factor 1A	12 ± 3	4 ± 1	14 ± 6	63 ± 7	86 ± 8
Q01852	TIM44, 356-370 (369)	Mit import inner memb. translocase subunit TIM44	10 ± 2	13 ± 4	16 ± 4	63 ± 13	78 ± 8
P23644	TOM40, 324-330 (326)	Mitochondrial import receptor subunit TOM40	17 ± 3	18 ± 5	32 ± 13	61 ± 11	75 ± 12
P00942	TPI1, 115-134 (126)	Triosephosphate isomerase	7 ± 4	9 ± 5	10 ± 2	56 ± 4	63 ± 6
P00942	TPI1, 27-55 (41)	Triosephosphate isomerase	7 ± 4	4 ± 2	5 ± 2	80 ± 8	93 ± 4
Q00764	TPS1, 206-222 (210)	Alpha,alpha-trehalose-phosphate synthase	6 ± 4	7 ± 2	10 ± 4	50 ± 7	76 ± 7
P00899	TRP2, 121-135 (133)	Anthranilate synthase component 1	6 ± 4	5 ± 4	7 ± 3	63 ± 3	87 ± 8
P00931	TRP5, 402-422 (420)	Tryptophan synthase	7 ± 3	13 ± 2	21 ± 7	96 ± 3	86 ± 12
P22515	UBA1, 596-603 (600)	Ubiquitin-activating enzyme E1 1	27 ± 3	22 ± 1	29 ± 7	55 ± 9	84 ± 7
P32861	UGP1, 351-369 (364)	UTP-glucose-1-phosphate uridylyltransferase	11 ± 3	9 ± 6	24 ± 13	71 ± 5	80 ± 4
P07259	URA2, 1832-1854 (1832)	Protein URA2	11 ± 4	17 ± 10	13 ± 7	73 ± 6	99 ± 0
P13298	URA5, 11-23 (19)	Orotate phosphoribosyltransferase 1	5 ± 3	13 ± 3	10 ± 6	74 ± 15	89 ± 8
P39968	VAC8, 341-359 (352)	Vacuolar protein 8	23 ± 5	20 ± 4	32 ± 5	64 ± 9	81 ± 16
P16140	VMA2, 170-189 (188)	V-type proton ATPase subunit B	17 ± 2	14 ± 4	21 ± 5	73 ± 13	82 ± 4
P32563	VPH1, 386-403 (394;396)	V-type proton ATPase subunit a	11 ± 1	8 ± 2	22 ± 4	62 ± 11	89 ± 5
P25491	YDJ1, 364-382 (370)	Mitochondrial import protein MAS5	7 ± 2	10 ± 6	9 ± 5	68 ± 12	82 ± 6
Q03034	YDR539W, 58-74 (69)	Uncharacterized protein YDR539W	15 ± 4	19 ± 5	22 ± 4	74 ± 7	90 ± 2
P53111	YGL157W, 232-247 (240)	Putative uncharact. oxidoreductase YGL157W	31 ± 6	19 ± 6	30 ± 3	70 ± 21	91 ± 3
P38708	YHR020W, 387-405 (396)	Putative prolyl-tRNA synthetase YHR020W	9 ± 4	4 ± 1	18 ± 7	61 ± 19	80 ± 7

Chapter 4: Collapse of the cellular redox balance in aging yeast   159

ORF	Gene	Description						
Q06252	YLR179C, 137-146 (142)	Uncharacterized protein YLR179C	4 ± 3	3 ± 2	13 ± 4	75 ± 13	91 ± 5
Q12305	YOR285W, 93-103 (98)	Putative thiosulfate sulfurtransferase YOR285W	21 ± 2	9 ± 6	18 ± 10	67 ± 1	79 ± 6
P41920	YRB1, 134-154 (135)	Ran-specific GTPase-activating protein 1	16 ± 6	9 ± 1	17 ± 4	69 ± 6	85 ± 5

B *accumulate in thiol-oxidized state at day 3 & oxidation at day 3 < 50%*

ORF	Gene	Description						
Q00955	ACC1, 1212-1241 (1225)	Acetyl-CoA carboxylase	16 ± 3	4 ± 2	19 ± 10	44 ± 6	66 ± 6
P46672	ARC1, 261-274 (266)	GU4 nucleic-binding protein 1	17 ± 9	14 ± 3	24 ± 6	45 ± 5	70 ± 8
P39077	CCT3, 512-531 (518)	T-complex protein 1 subunit gamma	15 ± 1	12 ± 6	18 ± 8	35 ± 3	39 ± 9
P42943	CCT7, 452-469 (454)	T-complex protein 1 subunit eta	11 ± 3	31 ± 7	16 ± 7	27 ± 4	59 ± 12
P25694	CDC48, 76-93 (92)	Cell division control protein 48	15 ± 8	14 ± 7	23 ± 8	42 ± 14	59 ± 3
P14832	CPR1, 36-53 (38)[a]	Peptidyl-prolyl cis-trans isomerase CPR1	12 ± 3	14 ± 0	19 ± 14	44 ± 2	63 ± 2
P32324	EFT2, 441-465 (448)	Elongation factor 2	8 ± 5	14 ± 8	17 ± 8	31 ± 9	33 ± 4
P47912	FAA4, 26-52 (49)	Long-chain-fatty-acid--CoA ligase 4	4 ± 3	24 ± 7	13 ± 5	33 ± 12	83 ± 11
P47912	FAA4, 400-414 (403)	Long-chain-fatty-acid--CoA ligase 4	9 ± 4	4 ± 1	14 ± 7	34 ± 15	76 ± 13
P14540	FBA1, 96-115 (112)	Fructose-bisphosphate aldolase	13 ± 2	5 ± 2	14 ± 3	30 ± 8	39 ± 6
P38625	GUA1, 438-455 (450)	GMP synthase	9 ± 3	18 ± 5	12 ± 8	30 ± 3	67 ± 7
P46655	GUS1, 378-400 (385)	Glutamyl-tRNA synthetase	15 ± 5	11 ± 4	25 ± 13	43 ± 8	64 ± 5
P00498	HIS1, 149-179 (160)	ATP phosphoribosyltransferase	9 ± 2	12 ± 2	18 ± 3	46 ± 12	69 ± 6
P31539	HSP104, 387-407 (400)	Heat shock protein 104	11 ± 4	7 ± 3	12 ± 2	28 ± 3	57 ± 6
P33416	HSP78, 325-337 (333)	Heat shock protein 78	13 ± 3	9 ± 4	15 ± 5	33 ± 1	65 ± 8
P04807	HXK2, 395-407 (398,404)	Hexokinase 2	17 ± 3	16 ± 4	25 ± 8	44 ± 4	57 ± 6
P25605	ILV6, 236-254 (244)	Acetolactate synthase small subunit	15 ± 3	9 ± 3	22 ± 7	40 ± 1	55 ± 1
P15180	KRS1, 206-229 (212)	Lysyl-tRNA synthetase	12 ± 6	8 ± 3	17 ± 5	24 ± 4	37 ± 3
Q12122	LYS21, 194-206 (202)	Homocitrate synthase	13 ± 4	13 ± 2	21 ± 7	32 ± 4	43 ± 3
Q12122	LYS21, 289-319 (314)	Homocitrate synthase	14 ± 6	20 ± 4	24 ± 3	46 ± 13	61 ± 9
P09440	MIS1, 216-229 (224)	C-1-tetrahydrofolate synthase	10 ± 3	14 ± 3	19 ± 6	35 ± 5	53 ± 5
P32379	PUP2, 94-122 (117)	Proteasome component PUP2	23 ± 10	15 ± 7	18 ± 10	42 ± 9	67 ± 7
Q12189	RKI1, 73-83 (75)	Ribose-5-phosphate isomerase	12 ± 5	11 ± 5	14 ± 3	16 ± 3	49 ± 8
Q3E757	RPL11B, 144-153 (144)	60S ribosomal protein L11-B	8 ± 5	11 ± 3	11 ± 7	23 ± 7	54 ± 4
P39741	RPL35B, 63-77 (67)	60S ribosomal protein L35	8 ± 4	11 ± 2	17 ± 12	27 ± 15	60 ± 11
P26321	RPL5, 55-85 (62)	60S ribosomal protein L5	13 ± 4	27 ± 5	17 ± 5	29 ± 7	75 ± 6
P32827	RPS23B, 83-109 (92)	40S ribosomal protein S23	8 ± 4	10 ± 1	15 ± 3	29 ± 8	34 ± 8
P05754	RPS8B, 179-195 (179)	40S ribosomal protein S8	14 ± 2	11 ± 6	14 ± 6	38 ± 17	50 ± 16
P29478	SEC65, 98-110 (104)	Signal recognition particle subunit SEC65	10 ± 4	10 ± 2	12 ± 6	37 ± 10	70 ± 6
P07284	SES1, 397-410 (400)	Seryl-tRNA synthetase	11 ± 2	14 ± 3	17 ± 12	42 ± 4	53 ± 8
P36024	SIS2, 2-23 (21)	Protein SIS2	18 ± 3	24 ± 4	21 ± 8	43 ± 10	38 ± 5
P11484	SSB1, 431-455 (435,454)	Heat shock protein SSB1	19 ± 3	18 ± 1	26 ± 13	42 ± 11	45 ± 6
P32589	SSE1, 223-235 (228)	Heat shock protein homolog SSE1	10 ± 3	16 ± 6	27 ± 4	22 ± 10	40 ± 5
P38353	SSH1, 70-77 (74)	Sec sixty-one protein homolog	7 ± 5	10 ± 5	18 ± 10	32 ± 7	82 ± 12
P36008	TEF4, 125-143 (129)	Elongation factor 1-gamma 2	7 ± 2	14 ± 7	8 ± 3	43 ± 16	62 ± 8
P39968	VAC8, 428-447 (445)	Vacuolar protein 8	12 ± 3	10 ± 1	12 ± 2	42 ± 17	54 ± 3

C *accumulate in thiol-oxidized state at day 3 & oxidation at log phase > 40%*

ORF	Gene	Description						
P00549	CDC19, 414-425 (418)	Pyruvate kinase 1	46 ± 7	39 ± 2	49 ± 12	76 ± 5	86 ± 7

UniProt	Gene, peptide (oxidation site)	Description					
P53849	GIS2, 67-78 (67;70)	Zinc finger protein GIS2	52 ± 6	39 ± 2	57 ± 16	79 ± 4	83 ± 7
Q99258	RIB3, 125-136 (133)	3,4-dihydroxy-2-butanone 4-phosphate synthase	45 ± 6	36 ± 4	53 ± 16	85 ± 5	97 ± 2
P51402	RPL37B, 33-43 (34;37)	60S ribosomal protein L37-B	40 ± 4	36 ± 10	40 ± 8	66 ± 8	75 ± 4
P39939	RPS26B, 67-82 (74;77)	40S ribosomal protein S26-B	46 ± 9	54 ± 12	55 ± 13	62 ± 4	87 ± 9
P41057	RPS29A, 23-32 (24)	40S ribosomal protein S29-A	47 ± 9	39 ± 8	53 ± 10	83 ± 14	90 ± 12
P41057	RPS29A, 33-40 (39)	40S ribosomal protein S29-A	46 ± 9	56 ± 8	59 ± 17	69 ± 16	88 ± 7
P41058	RPS29B, 33-40 (39)	40S ribosomal protein S29-B	55 ± 13	47 ± 6	63 ± 11	87 ± 8	92 ± 8
P05759	RPS31, 44-62 (45;50)	40S ribosomal protein S31	46 ± 4	42 ± 2	58 ± 14	83 ± 3	78 ± 2
P26783	RPS5, 85-92 (87)	40S ribosomal protein S5	40 ± 8	42 ± 12	54 ± 16	68 ± 18	73 ± 3
P39926	SSO2, 108-123 (122)	Protein SSO2	41 ± 4	34 ± 5	46 ± 4	59 ± 8	67 ± 12
D	**onset of oxidation precedes oxidation of cluster A-C by 48 hours**						
P39078	CCT4, 393-402 (399)	T-complex protein 1 subunit delta	12 ± 6	24 ± 8	58 ± 7	87 ± 5	86 ± 10
P25694	CDC48, 104-119 (115)	Cell division control protein 48	12 ± 3	29 ± 6	55 ± 7	75 ± 2	93 ± 4
P54839	ERG13, 289-301 (300)	Hydroxymethylglutaryl-CoA synthase	17 ± 6	38 ± 15	47 ± 6	80 ± 13	91 ± 6
Q05670	FUS2, 366-375 (371)	Nuclear fusion protein FUS2	18 ± 4	33 ± 10	57 ± 8	68 ± 4	80 ± 7
P14904	LAP4, 191-203 (202)	Vacuolar aminopeptidase 1	19 ± 3	38 ± 4	45 ± 8	83 ± 8	86 ± 9
P32327	PYC2, 211-222 (218)	Pyruvate carboxylase 2	11 ± 2	28 ± 1	46 ± 7	81 ± 13	93 ± 3
P49626	RPL4B, 85-95 (94)	60S ribosomal protein L4-B	11 ± 5	25 ± 14	34 ± 10	33 ± 13	92 ± 6
P02994	TEF1, 409-421 (409)	Elongation factor 1-alpha	13 ± 3	38 ± 7	40 ± 8	62 ± 11	70 ± 11
E	**onset of oxidation precedes oxidation of cluster A-C by 24 hours**						
P28777	ARO2, 210-224 (221)	Chorismate synthase	13 ± 1	13 ± 6	31 ± 15	71 ± 5	74 ± 10
P47079	CCT8, 335-343 (336)	T-complex protein 1 subunit theta	47 ± 10	40 ± 11	77 ± 7	76 ± 11	89 ± 8
P09950	HEM1, 386-391 (386)	5-aminolevulinate synthase	13 ± 4	21 ± 7	58 ± 15	77 ± 9	88 ± 12
P21954	IDP1, 393-403 (398)	Isocitrate dehydrogenase 1	18 ± 4	19 ± 3	41 ± 4	61 ± 13	85 ± 9
P20967	KGD1, 982-996 (983)	2-oxoglutarate dehydrogenase E1	25 ± 3	21 ± 5	49 ± 4	57 ± 11	84 ± 12
P07702	LYS2, 611-625 (614)	L-aminoadipate-semialdehyde dehydrogenase	15 ± 7	22 ± 7	49 ± 15	79 ± 14	86 ± 10
P00958	MES1, 353-366 (353)	Methionyl-tRNA synthetase	23 ± 2	34 ± 7	64 ± 9	86 ± 4	82 ± 9
P38219	OLA1, 43-62 (43)	Uncharacterized GTP-binding protein	17 ± 7	12 ± 2	46 ± 12	91 ± 3	83 ± 2
P09232	PRB1, 344-368 (361)	Cerevisin	18 ± 5	23 ± 10	57 ± 12	91 ± 10	94 ± 4
P07275	PUT2, 154-171 (162)	Delta-1-pyrroline-5-carboxylate dehydrogenase	12 ± 4	10 ± 6	35 ± 11	60 ± 6	66 ± 13
P02405	RPL42B, 68-76 (74)	60S ribosomal protein L42	15 ± 2	22 ± 6	49 ± 11	86 ± 6	86 ± 7
P26781	RPS11B, 58-67 (58)	40S ribosomal protein S11	18 ± 2	17 ± 1	37 ± 15	77 ± 6	81 ± 11
Q3E7Y3	RPS22B, 72-78 (72)	40S ribosomal protein S22-B	13 ± 4	9 ± 5	34 ± 7	62 ± 14	72 ± 16
P07284	SES1, 411-433 (413;414)	Seryl-tRNA synthetase	22 ± 4	23 ± 6	55 ± 6	66 ± 7	79 ± 4
P29509	TRR1, 138-153 (142;145)	Thioredoxin reductase 1	33 ± 6	33 ± 2	65 ± 3	77 ± 6	82 ± 1
P15731	UBC4, 103-129 (108)	Ubiquitin-conjugating enzyme E2 4	19 ± 4	23 ± 8	48 ± 5	70 ± 11	81 ± 13
P25654	YCR090C, 112-125 (124)	UPF0587 protein YCR090C	34 ± 8	39 ± 12	67 ± 4	76 ± 9	88 ± 9
P25491	YDJ1, 181-200 (185;188)	Mitochondrial protein import protein MAS5	46 ± 4	46 ± 4	82 ± 9	89 ± 8	75 ± 7
F	**remain reduced during the course of the experiment**						
P38077	ATP3, 115-126 (117)	ATP synthase subunit gamma	9 ± 4	15 ± 6	18 ± 9	20 ± 5	17 ± 6
P07258	CPA1, 8-27 (11)	Carbamoyl-phosphate synthase small chain	24 ± 7	30 ± 6	24 ± 11	32 ± 11	30 ± 4

Chapter 4: Collapse of the cellular redox balance in aging yeast   161

	Accession	Gene, positions	Description					
G	P39976	DLD3, 408-426 (409)	D-lactate dehydrogenase 3	13 ± 5	24 ± 10	25 ± 3	36 ± 4	20 ± 8
	P14020	DPM1, 168-175 (172)	Dolichol-phosphate mannosyltransferase	14 ± 4	10 ± 3	12 ± 3	17 ± 4	29 ± 4
	P07149	FAS1, 1296-1317 (1308)	Fatty acid synthase subunit beta	13 ± 2	18 ± 6	23 ± 9	21 ± 2	25 ± 3
	P18562	FUR1, 195-216 (215)	Uracil phosphoribosyltransferase	11 ± 6	11 ± 3	15 ± 4	16 ± 7	21 ± 4
	P11986	INO1, 19-29 (19)	Inositol-1-phosphate synthase	34 ± 4	39 ± 13	26 ± 6	33 ± 6	36 ± 3
	P49367	LYS4, 335-344 (340)	Homoaconitase	35 ± 5	23 ± 1	35 ± 6	49 ± 10	39 ± 1
	P38999	LYS9, 333-342 (340)	Saccharopine dehydrogenase	15 ± 4	5 ± 2	14 ± 3	16 ± 4	6 ± 2
	P32179	MET22, 339-355 (349)	3'(2'),5'-bisphosphate nucleotidase	21 ± 3	14 ± 2	13 ± 4	23 ± 6	22 ± 6
	P05694	MET6, 650-667 (657)	Methionine synthase	12 ± 2	9 ± 2	13 ± 4	20 ± 8	31 ± 9
	P36517	MRPL4, 62-76 (62)	54S ribosomal protein L4	9 ± 3	15 ± 4	15 ± 3	26 ± 4	24 ± 7
	P25293	NAP1, 231-250 (249)	Nucleosome assembly protein	16 ± 6	17 ± 8	19 ± 8	29 ± 5	35 ± 10
	P38325	OM14, 28-35 (29)	Mitochondrial outer membrane protein	4 ± 3	5 ± 3	28 ± 9	17 ± 4	15 ± 5
	P18239	PET9, 287-294 (288)	ADP,ATP carrier protein 2	13 ± 2	12 ± 6	18 ± 7	21 ± 1	22 ± 6
	P04451	RPL23B, 13-32 (25)	60S ribosomal protein L23	15 ± 4	24 ± 4	18 ± 9	16 ± 8	11 ± 6
	P38886	RPN10, 110-127 (115)	26S proteasome regulatory subunit RPN10	17 ± 2	18 ± 10	13 ± 5	14 ± 3	23 ± 7
	P10591	SSA1, 4-34 (15)	Heat shock protein SSA1	13 ± 3	13 ± 2	17 ± 4	25 ± 8	21 ± 2
	P53616	SUN4, 233-246 (239)	Septation protein SUN4	5 ± 2	11 ± 1	5 ± 3	9 ± 2	9 ± 2
	P07806	VAS1, 998-1031 (1015)	Valyl-tRNA synthetase	15 ± 7	15 ± 3	22 ± 3	26 ± 9	32 ± 6
	P32527	ZUO1, 162-178 (167)	Zuotin	13 ± 6	13 ± 4	12 ± 2	15 ± 4	19 ± 5
	remain oxidized during the course of the experiment							
	P50273	ATP22, 664-677 (672)	Mitochondrial translation factor ATP22	75 ± 6	64 ± 9	63 ± 11	58 ± 9	54 ± 3
	O13547	CCW14, 38-61 (42;51;53)	Covalently-linked cell wall protein 14	93 ± 3	96 ± 1	85 ± 5	88 ± 12	93 ± 2
	P20081	FPR1, 50-64 (55)	FK506-binding protein 1	76 ± 6	73 ± 6	70 ± 9	83 ± 9	79 ± 5
	P32472	FPR2, 30-39 (36)	FK506-binding protein 2	79 ± 3	73 ± 10	76 ± 2	83 ± 7	83 ± 12
	P53849	GIS2, 118-129 (118;121)	Zinc finger protein GIS2	70 ± 5	59 ± 4	67 ± 11	80 ± 5	78 ± 9
	P53849	GIS2, 48-59 (49;52)	Zinc finger protein GIS2	66 ± 7	70 ± 7	73 ± 6	76 ± 6	78 ± 7
	P25373	GRX1, 25-45 (27;30)	Glutaredoxin 1	80 ± 10	83 ± 9	91 ± 4	93 ± 6	97 ± 2
	P32486	KRE6, 476-493 (479;481)	Beta-glucan synthesis-associated protein	92 ± 3	90 ± 2	94 ± 2	95 ± 3	94 ± 4
	P14796	RPL40A, 13-21 (15;20)	60S ribosomal protein L40	66 ± 8	63 ± 15	79 ± 9	88 ± 3	79 ± 13
	P00445	SOD1, 138-154 (147)	Superoxide dismutase [Cu-Zn]	72 ± 7	63 ± 4	72 ± 14	80 ± 9	95 ± 3
	P00445	SOD1, 45-69 (58)	Superoxide dismutase [Cu-Zn]	69 ± 4	83 ± 12	84 ± 6	86 ± 7	95 ± 4
	P87108	TIM10, 44-64 (44;61)	Mit. import inner memb. translocase subunit TIM10	93 ± 5	99 ± 0	97 ± 2	95 ± 3	95 ± 2
	P15565	TRM1, 333-343 (333)	N(2),N(2)-dimethylguanos. tRNA methyltransferase	65 ± 13	68 ± 8	64 ± 10	63 ± 11	83 ± 13
	P25491	YDJ1, 140-149 (143;146)	Mitochondrial protein import protein MAS5	64 ± 5	53 ± 8	70 ± 2	68 ± 5	76 ± 5
	Q12512	ZPS1, 123-142 (123;130)	Protein ZPS1	96 ± 1	91 ± 5	91 ± 4	98 ± 1	94 ± 4
H	**minor increase in oxidation over several days**							
	P00330	ADH1, 277-299 (277;278)	Alcohol dehydrogenase 1	17 ± 5	32 ± 9	42 ± 7	45 ± 7	39 ± 8
	P32476	ERG1, 170-189 (174)	Squalene monooxygenase	27 ± 2	21 ± 6	46 ± 7	55 ± 15	73 ± 12
	P17709	GLK1, 440-460 (448)	Glucokinase 1	15 ± 6	13 ± 5	28 ± 5	42 ± 9	62 ± 5
	Q12068	GRE2, 77-105 (86)	NADPH-dependent methylglyoxal reductase	5 ± 2	13 ± 5	31 ± 6	29 ± 6	40 ± 4
	P19882	HSP60, 266-287 (283)	Heat shock protein 60	18 ± 3	14 ± 2	34 ± 4	31 ± 6	40 ± 2
	P36013	MAE1, 202-221 (220)	Malic enzyme	29 ± 2	20 ± 5	43 ± 9	43 ± 9	69 ± 10

Q12447	PAA1, 48-61 (51:55)	Polyamine N-acetyltransferase 1	26 ± 8	36 ± 10	53 ± 13	48 ± 12	61 ± 8
P40525	RPL34B, 44-60 (44;47)	60S ribosomal protein L34-B	43 ± 7	33 ± 5	52 ± 14	64 ± 9	79 ± 5
P14796	RPL40A, 8-17 (15)	60S ribosomal protein L40	26 ± 3	29 ± 4	44 ± 6	77 ± 7	84 ± 13
P29453	RPL8B, 150-171 (170)	60S ribosomal protein L8-B	9 ± 7	14 ± 4	26 ± 7	31 ± 10	69 ± 9
P41058	RPS29B, 23-32 (24)	40S ribosomal protein S29-B	41 ± 7	46 ± 7	66 ± 9	89 ± 3	98 ± 1
P32568	SNQ2, 1097-1121 (1098)	Protein SNQ2	30 ± 7	62 ± 10	42 ± 10	45 ± 15	86 ± 11
P38788	SSZ1, 65-85 (81)	Ribosome-associated complex subunit SSZ1	29 ± 8	27 ± 2	42 ± 12	58 ± 10	50 ± 7
P39968	VAC8, 300-317 (315)	Vacuolar protein 8	18 ± 4	16 ± 3	30 ± 17	36 ± 12	75 ± 5

Table 4.B. Thiol oxidation status of yeast proteins during chronological aging in 0.5% glucose CR media. OxICAT was applied to chronological aging yeast cells cultivated in CR media. Identified peptides are grouped according to their cluster (A-H). Gene designation, protein name, and Swiss-Prot accession numbers are based on annotations in SGD. "% oxidation" denotes the average oxidation of the redox-sensitive cysteine (second column, in brackets), obtained from three independent experiments ± standard deviation. *: peptide identical in both isoenzymes; [a,b,c]: at this mass, two different peptides were identified.

Cluster	Swiss-Prot	Gene, Peptide (Cys)	Protein name	log phase	Day 1	Day 2	Day 3	Day 4	Day 5	Day 7
A	\multicolumn{10}{l}{accumulate in thiol-oxidized state at day 5 & oxidation at day 5 > 50%}									
	P19414	ACO1, 376-392 (382)	Aconitase 1	7 ± 2	11 ± 7	11 ± 5	20 ± 8	28 ± 7	84 ± 5	88 ± 0
	P38009	ADE17, 285-304 (300)	Bifunctional purine biosynthesis protein	7 ± 0	9 ± 2	16 ± 1	18 ± 0	28 ± 7	66 ± 7	79 ± 5
	P07245	ADE3, 603-621 (612)	C-1-tetrahydrofolate synthase	22 ± 6	21 ± 6	29 ± 9	32 ± 5	47 ± 5	73 ± 8	62 ± 14
	P38013	AHP1, 114-124 (120)	Peroxiredoxin type-2	11 ± 4	6 ± 2	10 ± 3	13 ± 3	25 ± 10	76 ± 8	86 ± 2
	P38013	AHP1, 49-79 (62)	Peroxiredoxin type-2	16 ± 5	15 ± 3	17 ± 1	17 ± 2	22 ± 5	71 ± 4	93 ± 5
	P54115	ALD6, 124-134 (132)	Aldehyde dehydrogenase (Mg²⁺ dep.)	13 ± 4	13 ± 2	13 ± 2	16 ± 6	31 ± 3	83 ± 6	88 ± 4
	P54115	ALD6, 290-311 (306)	Aldehyde dehydrogenase (Mg²⁺ dep.)	25 ± 5	39 ± 8	21 ± 8	22 ± 2	26 ± 7	60 ± 10	88 ± 7
	P54783	ALO1, 359-374 (359)	D-arabinono-1,4-lactone oxidase	20 ± 5	25 ± 9	34 ± 7	22 ± 0	34 ± 7	68 ± 9	84 ± 4
	P32449	ARO4, 240-249 (244)	Phospho-2-dehydro-3-deoxyheptonate aldolase	18 ± 3	12 ± 2	23 ± 8	17 ± 2	18 ± 12	61 ± 4	79 ± 10
	P49089	ASN1, 421-440 (432)	Asparagine synthetase 1	7 ± 3	11 ± 4	15 ± 2	28 ± 3	30 ± 10	52 ± 7	60 ± 7
	P49089/P49090	ASN1/2, 396-406 (404)*	Asparagine synthetase 1/2	12 ± 2	12 ± 5	11 ± 3	14 ± 3	24 ± 4	63 ± 12	66 ± 12
	P38891	BAT1, 125-141 (126)	Branched-chain-amino-acid aminotransferase	10 ± 2	9 ± 4	9 ± 4	15 ± 8	26 ± 8	76 ± 3	90 ± 1
	P38891	BAT1, 220-237 (227)	Branched-chain-amino-acid aminotransferase	8 ± 0	10 ± 2	17 ± 7	14 ± 3	30 ± 6	60 ± 6	85 ± 6
	P29311	BMH1, 77-88 (85)	14-3-3 protein	10 ± 0	11 ± 7	14 ± 4	17 ± 4	32 ± 6	96 ± 4	90 ± 2
	P39077	CCT3, 512-531 (518)	T-complex protein 1 subunit gamma	11 ± 2	7 ± 3	13 ± 3	17 ± 9	34 ± 4	77 ± 13	81 ± 5
	P39079	CCT6, 515-535 (530)	T-complex protein 1 subunit zeta	11 ± 6	4 ± 2	4 ± 2	17 ± 3	27 ± 7	63 ± 4	66 ± 8
	P39079	CCT6, 79-103 (96)	T-complex protein 1 subunit zeta	9 ± 3	6 ± 1	9 ± 2	14 ± 7	27 ± 3	71 ± 2	94 ± 5
	P00549	CDC19, 370-394 (371)	Pyruvate kinase 1	28 ± 9	10 ± 3	13 ± 5	24 ± 5	30 ± 6	90 ± 3	86 ± 10
	P03965	CPA2, 159-182 (170)	Carbamoyl-phosphate synthase large chain	12 ± 0	14 ± 6	13 ± 2	14 ± 5	27 ± 6	78 ± 6	81 ± 7
	P14832	CPR1, 36-53 (38)[a]	Peptidyl-prolyl cis-trans isomerase CPR1	9 ± 2	6 ± 1	13 ± 3	17 ± 3	34 ± 3	71 ± 5	77 ± 4
	P53691	CPR6, 150-165 (156)	Peptidyl-prolyl cis-trans isomerase CPR6	51 ± 7	34 ± 7	43 ± 4	20 ± 7	27 ± 10	87 ± 7	93 ± 2
	P38845	CRP1, 105-118 (113)	Uncharacterized	25 ± 5	16 ± 8	20 ± 5	20 ± 3	39 ± 14	83 ± 7	85 ± 8
	P06634	DED1, 256-270 (257)	ATP-dependent RNA helicase DED1	15 ± 1	16 ± 3	9 ± 2	21 ± 6	27 ± 4	70 ± 5	95 ± 0
	P06634	DED1, 275-285 (276)	ATP-dependent RNA helicase DED1	11 ± 3	18 ± 7	10 ± 1	20 ± 1	19 ± 5	60 ± 5	77 ± 9
	P04802	DPS1, 250-259 (255)	Aspartyl-tRNA synthetase	29 ± 2	29 ± 1	31 ± 2	31 ± 2	35 ± 3	70 ± 11	81 ± 9
	Q06053	DUS3, 210-225 (224)	tRNA-dihydrouridine synthase 3	9 ± 1	14 ± 2	15 ± 7	23 ± 2	34 ± 5	70 ± 15	81 ± 12
	P32324	EFT2, 353-370 (366)[b]	Elongation factor 2	12 ± 4	5 ± 2	13 ± 6	14 ± 8	25 ± 4	73 ± 5	89 ± 4
	P32324	EFT2, 121-144 (136)	Elongation factor 2	13 ± 2	17 ± 4	17 ± 4	17 ± 6	30 ± 1	77 ± 10	83 ± 1
	P32324	EFT2, 724-749 (735)	Elongation factor 2	9 ± 2	8 ± 5	13 ± 1	20 ± 12	19 ± 3	70 ± 9	78 ± 8
	P00925	ENO2, 244-255 (248)	Enolase 2	8 ± 3	9 ± 3	10 ± 3	11 ± 7	19 ± 7	55 ± 6	82 ± 6

ID	Gene	Description							
P38333	ENP1, 320-333 (330)	Essential nuclear protein 1	9 ± 2	11 ± 1	10 ± 3	16 ± 4	22 ± 5	56 ± 13	79 ± 7
P41338	ERG10, 343-362 (358)[a]	Acetyl-CoA acetyltransferase	7 ± 2	6 ± 1	8 ± 2	15 ± 3	30 ± 2	76 ± 9	83 ± 2
P10614	ERG11, 137-146 (142)	Cytochrome P450	18 ± 3	15 ± 3	20 ± 8	17 ± 3	29 ± 8	69 ± 4	60 ± 8
P29704	ERG9, 340-350 (341)	Squalene synthetase	9 ± 2	12 ± 1	11 ± 4	21 ± 4	26 ± 5	69 ± 7	80 ± 6
P14540	FBA1, 250-264 (254)	Fructose-bisphosphate aldolase	11 ± 2	11 ± 3	15 ± 2	24 ± 2	44 ± 17	74 ± 10	81 ± 13
P32614	FRDS1, 446-459 (454)	Probable fumarate reductase	7 ± 2	16 ± 2	15 ± 3	27 ± 8	32 ± 12	66 ± 4	85 ± 7
P15624	FRS1, 121-138 (137)	Phenylalanyl-tRNA synthetase beta chain	9 ± 3	23 ± 10	11 ± 3	15 ± 5	30 ± 4	67 ± 2	75 ± 5
Q05670	FUS2, 366-375 (371)	Nuclear fusion protein FUS2	9 ± 4	18 ± 5	12 ± 4	20 ± 6	31 ± 3	64 ± 8	91 ± 2
P09032	GCD1, 163-188 (172)	Translation initiation factor eIF-2B subunit gamma	14 ± 5	13 ± 3	15 ± 2	26 ± 9	29 ± 7	72 ± 9	88 ± 5
P39958	GDI1, 217-226 (221)	Rab GDP-dissociation inhibitor	10 ± 3	13 ± 3	19 ± 7	29 ± 3	40 ± 10	61 ± 8	81 ± 10
P25370	GFD2, 88-103 (90)	Good for DBP5 activity protein 2	16 ± 4	14 ± 2	11 ± 4	16 ± 2	32 ± 10	96 ± 3	92 ± 5
P32598	GLC7, 20-40 (38)	Serine/threonine-protein phosphatase	8 ± 2	7 ± 3	12 ± 2	16 ± 4	30 ± 7	73 ± 1	88 ± 6
P41277/P40106	GPP1/GPP2, 210-222 (211;211)*	(DL)-glycerol-3-phosphatase 1/2	18 ± 8	24 ± 1	30 ± 4	27 ± 0	51 ± 14	75 ± 9	91 ± 3
P16622	HEM15, 118-136 (123)	Ferrochelatase	21 ± 7	13 ± 4	19 ± 4	24 ± 3	33 ± 8	62 ± 5	83 ± 12
P13663	HOM2, 235-249 (247)	Aspartate-semialdehyde dehydrogenase	11 ± 2	9 ± 2	15 ± 3	14 ± 3	27 ± 4	54 ± 7	86 ± 8
P13663	HOM2, 250-263 (259)	Aspartate-semialdehyde dehydrogenase	19 ± 1	13 ± 1	17 ± 2	24 ± 2	29 ± 4	57 ± 10	73 ± 7
P10869	HOM3, 38-56 (53)	Aspartokinase	6 ± 1	16 ± 9	23 ± 4	18 ± 8	33 ± 11	57 ± 2	79 ± 6
P04807	HXK2, 395-407 (398;404)	Hexokinase 2	17 ± 3	15 ± 2	22 ± 4	30 ± 5	38 ± 6	68 ± 2	89 ± 8
P21954	IDP1, 89-98 (89)	Isocitrate dehydrogenase 1	19 ± 3	17 ± 5	19 ± 1	25 ± 4	44 ± 5	70 ± 14	84 ± 5
P39522	ILV3, 157-180 (176)	Dihydroxy-acid dehydratase	22 ± 9	12 ± 2	18 ± 6	27 ± 3	32 ± 7	77 ± 8	83 ± 5
P06168	ILV5, 307-321 (316)	Keto-acid reductoisomerase	18 ± 4	9 ± 2	20 ± 4	14 ± 2	26 ± 10	90 ± 2	87 ± 7
P25605	ILV6, 102-120 (111)	Acetolactate synthase small subunit	9 ± 1	8 ± 3	7 ± 3	14 ± 4	25 ± 5	85 ± 3	86 ± 9
P25605	ILV6, 261-272 (270)	Acetolactate synthase small subunit	8 ± 2	10 ± 6	13 ± 2	13 ± 3	23 ± 4	65 ± 5	73 ± 10
P25605	ILV6, 76-93 (82)	Acetolactate synthase small subunit	7 ± 2	7 ± 2	9 ± 4	11 ± 2	18 ± 9	61 ± 7	84 ± 15
P40069	KAP123, 977-987 (984)	Importin subunit beta-4	19 ± 7	18 ± 4	20 ± 9	23 ± 8	17 ± 6	55 ± 4	83 ± 10
P15180	KRS1, 429-446 (438)	Lysyl-tRNA synthetase	9 ± 1	17 ± 3	11 ± 4	15 ± 4	27 ± 6	66 ± 5	81 ± 8
P15180	KRS1, 430-446 (438)	Lysyl-tRNA synthetase	9 ± 2	26 ± 6	20 ± 12	28 ± 9	36 ± 9	79 ± 10	83 ± 3
P14904	LAP4, 191-203 (202)	Vacuolar aminopeptidase 1	25 ± 3	35 ± 9	27 ± 5	22 ± 2	39 ± 6	61 ± 2	81 ± 2
P07264	LEU1, 60-77 (63)	3-isopropylmalate dehydratase	13 ± 3	15 ± 2	23 ± 4	22 ± 3	42 ± 7	87 ± 7	85 ± 1
P06208	LEU4, 345-353 (345)	2-isopropylmalate synthase	8 ± 3	9 ± 2	11 ± 3	14 ± 5	26 ± 12	65 ± 12	80 ± 7
P09624	LPD1, 45-58 (54)	Dihydrolipoyl dehydrogenase	15 ± 1	18 ± 3	18 ± 6	29 ± 2	35 ± 13	73 ± 5	88 ± 4
P36013	MAE1, 535-555 (541)	Malic enzyme	10 ± 2	15 ± 1	19 ± 2	19 ± 2	17 ± 7	71 ± 4	90 ± 4
P53091	MCM6, 813-832 (815)	DNA replication licensing factor MCM6	5 ± 3	18 ± 6	13 ± 3	16 ± 4	27 ± 5	71 ± 9	88 ± 9
P53875	MRPL19, 36-46 (40)[c]	54S ribosomal protein L19	9 ± 5	9 ± 3	8 ± 1	14 ± 2	25 ± 5	75 ± 1	77 ± 11
P53081	NIF3, 105-123 (114)	NGG1-interacting factor 3	12 ± 1	9 ± 2	14 ± 2	16 ± 4	24 ± 8	51 ± 10	69 ± 6
P15646	NOP1, 273-288 (275)	rRNA 2'-O-methyltransferase fibrillarin	13 ± 3	14 ± 3	11 ± 5	16 ± 8	29 ± 9	69 ± 6	83 ± 11
P15646	NOP1, 312-319 (314)	rRNA 2'-O-methyltransferase fibrillarin	10 ± 3	14 ± 8	9 ± 4	21 ± 1	36 ± 8	71 ± 6	81 ± 4
P27476	NSR1, 233-254 (241)	Nuclear localization sequence-binding protein	12 ± 4	22 ± 4	30 ± 12	32 ± 5	34 ± 12	67 ± 10	79 ± 6
P38219	OLA1, 126-143 (126)	Uncharacterized GTP-binding protein	6 ± 2	8 ± 3	6 ± 2	13 ± 2	25 ± 6	66 ± 6	81 ± 6
P38219	OLA1, 72-81 (76)	Uncharacterized GTP-binding protein	6 ± 2	16 ± 9	15 ± 2	20 ± 4	26 ± 6	65 ± 11	85 ± 3
P16467	PDC5, 128-154 (152)	Pyruvate decarboxylase isozyme 2	14 ± 3	10 ± 4	14 ± 2	16 ± 8	30 ± 8	77 ± 9	88 ± 4
P18239	PET9, 66-76 (73)[c]	ADP,ATP carrier protein 2	11 ± 5	4 ± 1	12 ± 1	17 ± 3	33 ± 12	76 ± 4	66 ± 9
P18239	PET9, 265-273 (271)	ADP,ATP carrier protein 2	8 ± 2	10 ± 5	10 ± 2	14 ± 3	26 ± 3	57 ± 11	81 ± 8
P18239	PET9, 67-76 (73)	ADP,ATP carrier protein 2	9 ± 2	10 ± 2	11 ± 4	14 ± 4	29 ± 2	78 ± 4	86 ± 11

ID	Gene, position	Description							
P07274	PFY1, 77-91 (89)	Profilin	5 ± 2	7 ± 4	14 ± 2	13 ± 3	24 ± 7	73 ± 10	94 ± 5
P00560	PGK1, 83-108 (98)	Phosphoglycerate kinase	11 ± 1	8 ± 1	10 ± 2	12 ± 0	23 ± 6	75 ± 4	88 ± 5
P37012	PGM2, 372-389 (376)[b]	Phosphoglucomutase-2	10 ± 2	7 ± 2	12 ± 1	10 ± 4	11 ± 3	63 ± 10	85 ± 9
P05030/P19657	PMA1/2, 392-414 (376,405)*	Plasma membrane ATPase 1/2	13 ± 1	12 ± 2	10 ± 4	10 ± 2	19 ± 6	54 ± 5	73 ± 6
P05030/P19657	PMA1/2, 415-443 (409,438)*	Plasma membrane ATPase 1/2	10 ± 1	6 ± 3	12 ± 4	11 ± 3	19 ± 8	59 ± 7	83 ± 8
P05030/P19657	PMA1/2, 487-503 (472,501)*	Plasma membrane ATPase 1/2	17 ± 2	14 ± 2	11 ± 2	14 ± 3	26 ± 3	87 ± 6	81 ± 10
P04840	POR1, 125-132 (130)	Mitochondrial outer membrane protein porin 1	7 ± 2	7 ± 1	11 ± 2	14 ± 0	28 ± 7	55 ± 10	70 ± 4
P21242	PRE10, 73-86 (76)	Proteasome component C1	17 ± 4	15 ± 7	19 ± 12	22 ± 8	29 ± 5	77 ± 3	80 ± 7
P07257	QCR2, 129-152 (137)	Cytochrome b-c1 complex subunit 2	20 ± 5	24 ± 8	15 ± 6	18 ± 4	35 ± 7	88 ± 4	81 ± 4
P32628	RAD23, 364-371 (365)	UV excision repair protein RAD23	17 ± 1	25 ± 4	16 ± 7	27 ± 0	35 ± 6	70 ± 9	90 ± 6
P17745	RNA1, 169-182 (180)	Ran GTPase-activating protein 1	9 ± 2	15 ± 4	26 ± 9	22 ± 7	36 ± 4	82 ± 8	81 ± 2
P49723	RNR4, 221-235 (228)	Ribonucleoside-diphosphate reductase small chain 2	12 ± 1	16 ± 2	13 ± 5	15 ± 5	22 ± 11	68 ± 6	92 ± 3
P41805	RPL10, 70-82 (71)	60S ribosomal protein L10	18 ± 2	13 ± 3	16 ± 2	26 ± 8	31 ± 4	60 ± 12	86 ± 9
P17079	RPL12B, 131-142 (141)	60S ribosomal protein L12	10 ± 2	20 ± 8	21 ± 7	15 ± 5	26 ± 9	63 ± 11	78 ± 6
P14796	RPL40A, 39-48 (39)	60S ribosomal protein L40	18 ± 5	19 ± 1	31 ± 4	20 ± 5	38 ± 10	69 ± 4	81 ± 2
P14796	RPL40A, 8-17 (15)	60S ribosomal protein L40	27 ± 2	27 ± 6	45 ± 6	33 ± 7	41 ± 8	84 ± 4	84 ± 5
P26781	RPS11B, 117-133 (128)	40S ribosomal protein S11	15 ± 2	17 ± 1	15 ± 3	24 ± 10	19 ± 4	77 ± 4	74 ± 7
P26781	RPS11B, 58-67 (58)	40S ribosomal protein S11	14 ± 2	17 ± 5	27 ± 6	25 ± 5	36 ± 11	83 ± 9	86 ± 2
P39516	RPS14B, 72-83 (72)	40S ribosomal protein S14-B	10 ± 5	18 ± 2	16 ± 3	22 ± 7	32 ± 9	56 ± 13	89 ± 8
P14127	RPS17B, 34-45 (35)	40S ribosomal protein S17-B	8 ± 2	6 ± 1	12 ± 2	15 ± 3	25 ± 5	84 ± 6	88 ± 3
P23248	RPS1B, 65-83 (69)	40S ribosomal protein S1-B	5 ± 1	7 ± 2	13 ± 2	15 ± 4	28 ± 5	68 ± 1	89 ± 7
Q3E7Y3	RPS22B, 72-78 (72)	40S ribosomal protein S22-B	32 ± 9	29 ± 8	36 ± 9	32 ± 11	41 ± 10	72 ± 14	79 ± 8
P05759	RPS37, 68-73 (68)	40S ribosomal protein S31	20 ± 6	21 ± 7	35 ± 9	25 ± 2	23 ± 6	66 ± 8	77 ± 4
P05754	RPS8B, 165-178 (168)	40S ribosomal protein S8	10 ± 2	16 ± 6	19 ± 11	25 ± 9	30 ± 8	62 ± 3	74 ± 4
P39954	SAH1, 192-199 (198)	Adenosylhomocysteinase	19 ± 3	22 ± 9	19 ± 19	21 ± 0	40 ± 5	75 ± 7	70 ± 14
P39954	SAH1, 218-236 (231)	Adenosylhomocysteinase	8 ± 5	16 ± 11	17 ± 9	26 ± 12	33 ± 3	76 ± 7	90 ± 3
P21243	SCL1, 63-77 (74)	Proteasome component C7-alpha	14 ± 4	23 ± 7	13 ± 7	15 ± 1	28 ± 9	60 ± 14	85 ± 6
P47052	SDH1B, 613-632 (624)	Succinate dehydrogenase flavoprotein subunit 2	13 ± 0	12 ± 5	19 ± 3	19 ± 6	36 ± 8	87 ± 6	85 ± 5
P29478	SEC65, 98-110 (104)	Signal recognition particle subunit SEC65	12 ± 2	13 ± 7	8 ± 2	16 ± 4	31 ± 4	80 ± 10	77 ± 11
P33330	SER1, 182-202 (182)	Phosphoserine aminotransferase	8 ± 3	8 ± 3	13 ± 4	14 ± 2	25 ± 8	60 ± 9	72 ± 11
P07284	SES1, 363-379 (370,373)	Seryl-tRNA synthetase	21 ± 2	29 ± 5	36 ± 2	32 ± 9	42 ± 13	58 ± 6	74 ± 0
P07284	SES1, 397-410 (400)	Seryl-tRNA synthetase	9 ± 0	12 ± 2	12 ± 3	10 ± 5	25 ± 10	67 ± 0	77 ± 8
P37292	SHM1, 89-106 (103)	Serine hydroxymethyltransferase	13 ± 2	11 ± 2	15 ± 2	14 ± 4	27 ± 3	79 ± 10	88 ± 2
P34223	SHP1, 78-103 (88)	UBX domain-containing protein 1	14 ± 4	16 ± 3	16 ± 7	17 ± 3	33 ± 5	66 ± 10	78 ± 6
P10592	SSA2, 304-313 (308)	Heat shock protein SSA2	10 ± 5	8 ± 1	13 ± 2	12 ± 4	21 ± 8	73 ± 9	90 ± 7
P11484	SSB1, 431-451 (435)	Heat shock protein SSB1	13 ± 1	7 ± 3	17 ± 3	16 ± 8	24 ± 9	79 ± 14	80 ± 9
P40150	SSB2, 452-466 (454)	Heat shock protein SSB2	19 ± 8	19 ± 7	16 ± 6	22 ± 7	31 ± 5	51 ± 9	81 ± 5
P32590	SSE2, 375-388 (380)	Heat shock protein homolog SSE2	6 ± 2	10 ± 4	11 ± 2	11 ± 5	19 ± 11	71 ± 6	89 ± 3
Q07478	SUB2, 359-368 (360)	ATP-dependent RNA helicase SUB2	11 ± 8	21 ± 7	22 ± 3	20 ± 3	34 ± 4	83 ± 4	83 ± 2
P00359/P00358	TDH2/3, 144-160 (150,154)*	Glyceraldehyde-3-phosphate dehydrogenase 2/3	24 ± 3	22 ± 4	21 ± 6	26 ± 4	28 ± 5	72 ± 8	87 ± 7
P02994	TEF1, 21-37 (31)	Elongation factor 1-alpha	15 ± 3	19 ± 4	15 ± 1	14 ± 5	29 ± 4	81 ± 1	88 ± 2
P02994	TEF1, 320-328 (324)	Elongation factor 1-alpha	13 ± 1	14 ± 6	16 ± 2	21 ± 5	22 ± 11	67 ± 11	84 ± 2
P02994	TEF1, 409-421 (409)	Elongation factor 1-alpha	18 ± 3	30 ± 4	17 ± 4	26 ± 6	36 ± 4	57 ± 13	68 ± 3
P16521	TEF3, 708-732 (731)	Elongation factor 3A	12 ± 2	14 ± 5	9 ± 2	17 ± 7	20 ± 9	80 ± 1	81 ± 7

ID	Protein, peptide (site)	Description							
P17255	TFP1, 356-367 (358)	V-type proton ATPase catalytic subunit A	10 ± 3	6 ± 3	18 ± 8	16 ± 2	30 ± 7	73 ± 4	89 ± 3
P17423	THR1, 216-231 (216)	Homoserine kinase	8 ± 3	12 ± 1	18 ± 11	29 ± 9	36 ± 6	79 ± 14	86 ± 0
P16120	THR4, 416-432 (428)	Threonine synthase	12 ± 2	8 ± 3	14 ± 1	23 ± 12	22 ± 5	66 ± 2	70 ± 3
P38912	TIF11, 83-101 (89)	Eukaryotic translation initiation factor 1A	13 ± 1	9 ± 3	11 ± 5	14 ± 2	22 ± 5	82 ± 13	86 ± 4
Q01852	TIM44, 356-370 (369)	Mit. import inner memb. translocase subunit TIM44	8 ± 2	18 ± 4	15 ± 4	27 ± 5	29 ± 10	67 ± 11	95 ± 3
P23644	TOM40, 324-330 (326)	Mitochondrial import receptor subunit TOM40	16 ± 4	19 ± 2	30 ± 8	16 ± 4	32 ± 3	57 ± 7	78 ± 8
P00942	TPI1, 27-55 (41)	Triosephosphate isomerase	5 ± 2	3 ± 1	8 ± 2	14 ± 8	27 ± 4	74 ± 8	90 ± 3
P00899	TRP2, 121-135 (133)	Anthranilate synthase component 1	7 ± 3	7 ± 2	14 ± 2	16 ± 3	25 ± 3	63 ± 0	86 ± 7
P00931	TRP5, 402-422 (420)	Tryptophan synthase	15 ± 2	23 ± 9	30 ± 3	29 ± 3	25 ± 5	79 ± 5	91 ± 3
P07259	URA2, 1832-1854 (1832)	Protein URA2	12 ± 1	11 ± 1	9 ± 1	20 ± 3	37 ± 11	60 ± 7	89 ± 9
P13298	URA5, 11-23 (19)	Orotate phosphoribosyltransferase 1	6 ± 2	7 ± 4	13 ± 4	21 ± 2	26 ± 4	71 ± 8	60 ± 9
P39968	VAC8, 300-317 (315)	Vacuolar protein 8	17 ± 2	15 ± 1	15 ± 4	17 ± 2	23 ± 11	53 ± 14	67 ± 10
P39968	VAC8, 341-359 (352)	Vacuolar protein 8	14 ± 7	17 ± 7	19 ± 2	14 ± 2	27 ± 8	62 ± 5	83 ± 6
P16140	VMA2, 170-189 (188)	V-type proton ATPase subunit B	14 ± 1	12 ± 4	14 ± 1	15 ± 3	30 ± 2	66 ± 14	79 ± 7
P32563	VPH1, 386-403 (394,396)	V-type proton ATPase subunit a	12 ± 2	10 ± 2	21 ± 2	18 ± 4	23 ± 6	66 ± 7	67 ± 7
P25491	YDJ1, 364-382 (370)	Mitochondrial protein import protein MAS5	8 ± 2	16 ± 3	11 ± 3	13 ± 6	25 ± 5	75 ± 6	77 ± 10
Q03034	YDR539W, 58-74 (69)	Uncharacterized protein YDR539W	15 ± 0	21 ± 5	27 ± 6	31 ± 4	51 ± 10	63 ± 9	88 ± 7
P38708	YHR020W, 387-405 (396)	Putative prolyl-tRNA synthetase YHR020W	13 ± 0	19 ± 4	7 ± 1	15 ± 3	25 ± 7	70 ± 3	81 ± 4
Q12305	YOR285W, 93-103 (98)	Putative thiosulfate sulfurtransferase YOR285W	18 ± 1	17 ± 3	13 ± 2	17 ± 2	32 ± 6	66 ± 5	74 ± 10
P41920	YRB1, 134-154 (135)	Ran-specific GTPase-activating protein 1	16 ± 3	12 ± 0	16 ± 6	14 ± 4	27 ± 8	60 ± 6	76 ± 13
B	*accumulate in thiol-oxidized state at day 5 & oxidation at day 5 < 50%*								
P16550	APA1, 127-139 (129)	AP4A phosphorylase	8 ± 4	15 ± 6	7 ± 2	12 ± 5	18 ± 8	9 ± 2	70 ± 12
P46672	ARC1, 261-274 (266)	GU4 nucleic-binding protein 1	10 ± 3	20 ± 3	19 ± 6	19 ± 5	14 ± 7	25 ± 3	61 ± 4
P42943	CCT7, 452-469 (454)	T-complex protein 1 subunit eta	7 ± 4	20 ± 4	13 ± 3	19 ± 2	32 ± 7	40 ± 4	88 ± 5
P25694	CDC48, 76-93 (92)	Cell division control protein 48	13 ± 3	12 ± 3	18 ± 6	24 ± 7	36 ± 17	39 ± 2	80 ± 7
P47912	FAA1, 26-52 (49)	Long-chain-fatty-acid--CoA ligase 4	11 ± 4	15 ± 1	13 ± 3	22 ± 4	30 ± 4	26 ± 5	55 ± 4
P47912	FAA4, 400-414 (403)	Long-chain-fatty-acid--CoA ligase 4	8 ± 2	10 ± 5	10 ± 3	15 ± 3	19 ± 5	33 ± 7	68 ± 7
P17709	GLK1, 440-460 (448)	Glucokinase 1	11 ± 3	10 ± 6	11 ± 3	14 ± 4	27 ± 7	36 ± 1	70 ± 4
P38720	GND1, 287-295 (287)	6-phosphogluconate dehydrogenase	15 ± 3	14 ± 2	18 ± 1	24 ± 11	24 ± 11	39 ± 12	56 ± 12
P38625	GUA1, 438-455 (450)	GMP synthase	8 ± 2	12 ± 1	14 ± 1	15 ± 6	27 ± 3	42 ± 8	77 ± 10
P46655	GUS1, 378-400 (385)	Glutamyl-tRNA synthetase	11 ± 4	12 ± 6	14 ± 4	17 ± 8	23 ± 11	31 ± 5	56 ± 8
P31539	HSP104, 387-407 (400)	Heat shock protein 104	11 ± 2	5 ± 1	9 ± 3	14 ± 7	32 ± 3	36 ± 9	87 ± 6
P21954	IDP1, 360-371 (370)	Isocitrate dehydrogenase 1	17 ± 4	12 ± 6	24 ± 4	29 ± 8	13 ± 6	34 ± 11	57 ± 0
P25605	ILV6, 236-254 (244)	Acetolactate synthase small subunit	7 ± 3	14 ± 2	16 ± 6	23 ± 9	25 ± 5	41 ± 5	69 ± 5
P07264	LEU1, 230-242 (233)	3-isopropylmalate dehydratase	13 ± 3	9 ± 5	14 ± 2	16 ± 7	15 ± 2	27 ± 6	64 ± 10
P09440	MIS1, 216-229 (224)	C-1-tetrahydrofolate synthase	5 ± 0	11 ± 6	12 ± 3	16 ± 7	32 ± 7	44 ± 3	60 ± 10
P32379	PUP2, 94-122 (117)	Proteasome component PUP2	20 ± 6	19 ± 1	17 ± 2	27 ± 3	30 ± 9	47 ± 4	79 ± 4
Q3E757	RPL11B, 144-153 (144)	60S ribosomal protein L11-B	14 ± 6	14 ± 7	8 ± 3	18 ± 2	24 ± 7	41 ± 9	62 ± 12
Q3E754	RPS21B, 16-22 (17)	40S ribosomal protein S21-B	16 ± 9	14 ± 1	16 ± 7	21 ± 10	22 ± 6	47 ± 4	57 ± 9
P05754	RPS8B, 179-195 (179)	40S ribosomal protein S8	15 ± 3	11 ± 2	13 ± 4	17 ± 5	30 ± 9	44 ± 10	71 ± 12
Q12464	RVB2, 221-232 (224)	RuvB-like protein 2	15 ± 3	22 ± 11	13 ± 3	17 ± 1	30 ± 6	39 ± 5	64 ± 9
P37291	SHM2, 188-198 (197)	Serine hydroxymethyltransferase	15 ± 5	3 ± 1	33 ± 2	18 ± 5	24 ± 5	44 ± 5	78 ± 11
P10592	SSA2, 14-34 (15)	Heat shock protein SSA2	13 ± 3	11 ± 1	14 ± 6	22 ± 3	20 ± 9	47 ± 9	81 ± 7

ID	Gene	Description								
P38353	SSH1, 70-77 (74)	Sec sixty-one protein homolog	6 ± 4	7 ± 4	13 ± 6	15 ± 4	27 ± 9	40 ± 5	87 ± 8
P16521	TEF3, 665-696 (691)	Elongation factor 3A	17 ± 5	16 ± 7	20 ± 7	15 ± 3	31 ± 4	49 ± 5	73 ± 8
P36008	TEF4, 125-143 (129)	Elongation factor 1-gamma 2	9 ± 2	8 ± 4	12 ± 3	14 ± 2	26 ± 8	48 ± 6	61 ± 5
P00942	TPI1, 115-134 (126)	Triosephosphate isomerase	4 ± 3	5 ± 1	5 ± 2	10 ± 5	10 ± 4	30 ± 13	78 ± 4
Q00764	TPS1, 206-222 (210)	Alpha,alpha-trehalose-phosphate synthase	12 ± 4	30 ± 10	13 ± 4	19 ± 5	21 ± 4	45 ± 5	76 ± 12
P22515	UBA1, 596-603 (600)	Ubiquitin-activating enzyme E1 1	14 ± 3	17 ± 1	27 ± 2	15 ± 8	31 ± 12	20 ± 6	54 ± 5
P39968	VAC8, 428-447 (445)	Vacuolar protein 8	11 ± 4	11 ± 5	7 ± 2	17 ± 7	34 ± 6	45 ± 5	73 ± 5

C   *accumulate in thiol-oxidized state at day 5 & oxidation at log phase > 40%*

ID	Gene	Description								
P00549	CDC19, 414-425 (418)	Pyruvate kinase 1	41 ± 5	42 ± 4	35 ± 1	36 ± 0	46 ± 7	72 ± 10	83 ± 7
P27614	CPS1, 361-370 (368)	Carboxypeptidase S	40 ± 4	40 ± 8	46 ± 1	37 ± 10	43 ± 3	62 ± 2	74 ± 4
P40525	RPL34B, 44-60 (44,47)	60S ribosomal protein L34-B	40 ± 3	33 ± 7	51 ± 2	43 ± 12	63 ± 12	67 ± 3	88 ± 3
P39939	RPS26B, 67-82 (74;77)	40S ribosomal protein S26-B	51 ± 3	50 ± 1	64 ± 7	59 ± 11	64 ± 3	63 ± 5	78 ± 9
P41057	RPS29A, 23-32 (24)	40S ribosomal protein S29-A	49 ± 2	42 ± 9	58 ± 3	57 ± 2	60 ± 3	78 ± 4	85 ± 4
P41057	RPS29A, 33-40 (39)	40S ribosomal protein S29-A	45 ± 3	47 ± 6	55 ± 8	59 ± 6	71 ± 2	88 ± 6	88 ± 4
P41058	RPS29B, 23-32 (24)	40S ribosomal protein S29-B	45 ± 4	36 ± 5	50 ± 13	46 ± 11	62 ± 3	87 ± 4	83 ± 7
P05759	RPS31, 44-62 (45,50)	40S ribosomal protein S31	41 ± 4	45 ± 10	36 ± 10	50 ± 5	57 ± 8	64 ± 9	84 ± 5
P00445	SOD1, 138-154 (147)	Superoxide dismutase [Cu-Zn]	52 ± 3	47 ± 8	56 ± 7	60 ± 10	58 ± 4	83 ± 4	93 ± 4
P39926	SSO2, 108-123 (122)	Protein SSO2	44 ± 5	40 ± 9	28 ± 7	32 ± 4	46 ± 2	59 ± 8	74 ± 13

D   *onset of oxidation precedes oxidation of cluster A-C by 48 hours*

ID	Gene	Description								
P19097	FAS2, 900-919 (917)	Fatty acid synthase subunit alpha	14 ± 5	8 ± 4	15 ± 3	59 ± 3	81 ± 2	87 ± 8	72 ± 10
P07702	LYS2, 611-625 (614)	L-aminoadipate-semialdehyde dehydrogenase	14 ± 4	11 ± 3	12 ± 5	48 ± 8	56 ± 3	74 ± 9	81 ± 6
P00958	MES1, 353-366 (353)	Methionyl-tRNA synthetase	20 ± 0	21 ± 7	31 ± 5	61 ± 7	78 ± 3	80 ± 7	98 ± 1
Q12447	PAA1, 48-61 (51;55)	Polyamine N-acetyltransferase 1	18 ± 6	15 ± 5	25 ± 8	52 ± 3	61 ± 4	71 ± 5	83 ± 1
P32327	PYC2, 211-222 (218)	Pyruvate carboxylase 2	13 ± 7	11 ± 4	22 ± 10	47 ± 10	66 ± 12	77 ± 3	81 ± 7
P39741	RPL35B, 63-77 (67)	60S ribosomal protein L35	16 ± 1	6 ± 2	17 ± 7	47 ± 3	48 ± 4	75 ± 6	93 ± 3
P02405	RPL42B, 68-76 (74)	60S ribosomal protein L42	15 ± 4	19 ± 6	25 ± 5	49 ± 9	54 ± 5	77 ± 10	93 ± 4
P07284	SES1, 411-433 (413;414)	Seryl-tRNA synthetase	20 ± 5	19 ± 5	17 ± 6	43 ± 7	65 ± 4	71 ± 7	81 ± 7
P32568	SNQ2, 1097-1121 (1098)	Protein SNQ2	38 ± 12	23 ± 5	38 ± 12	63 ± 8	67 ± 6	76 ± 5	84 ± 10

E   *onset of oxidation precedes oxidation of cluster A-C by 24 hours*

ID	Gene	Description								
P47143	ADO1, 121-130 (121)	Adenosine kinase	12 ± 0	13 ± 2	9 ± 2	13 ± 2	39 ± 4	79 ± 5	80 ± 10
P39078	CCT4, 393-402 (399)	T-complex protein 1 subunit delta	12 ± 2	14 ± 3	13 ± 2	11 ± 3	39 ± 8	72 ± 12	82 ± 13
P25694	CDC48, 104-119 (115)	Cell division control protein 48	13 ± 2	15 ± 9	14 ± 6	28 ± 8	62 ± 6	68 ± 4	90 ± 5
P25379	CHA1, 175-195 (178)	L-serine/threonine dehydratase	9 ± 3	9 ± 4	12 ± 2	15 ± 3	43 ± 6	70 ± 13	81 ± 11
P33892	GCN1, 1070-1078 (1077)	Translational activator GCN1	15 ± 3	16 ± 4	18 ± 3	24 ± 10	48 ± 9	47 ± 10	63 ± 11
P21954	IDP1, 393-403 (398)	Isocitrate dehydrogenase 1	16 ± 0	18 ± 6	32 ± 10	25 ± 5	49 ± 6	81 ± 9	80 ± 8
P20967	KGD1, 982-996 (983)	2-oxoglutarate dehydrogenase E1	18 ± 3	18 ± 8	19 ± 9	16 ± 6	57 ± 8	54 ± 6	70 ± 4
P38219	OLA1, 43-62 (43)	Uncharacterized GTP-binding protein	20 ± 9	10 ± 3	22 ± 9	29 ± 3	53 ± 2	78 ± 8	88 ± 8
P09232	PRB1, 344-368 (361)	Cerevisin	12 ± 3	14 ± 4	26 ± 4	23 ± 4	77 ± 7	80 ± 11	88 ± 8
Q99258	RIB3, 125-136 (133)	3,4-dihydroxy-2-butanone 4-phosphate synthase	37 ± 2	29 ± 9	37 ± 11	37 ± 3	59 ± 3	61 ± 6	86 ± 6
P41805	RPL10, 41-69 (49)	60S ribosomal protein L10	21 ± 3	17 ± 4	28 ± 8	23 ± 3	59 ± 7	80 ± 12	90 ± 7
P51402	RPL37B, 33-43 (34;37)	60S ribosomal protein L37-B	38 ± 6	34 ± 6	29 ± 10	39 ± 6	65 ± 4	78 ± 7	88 ± 9

P46654	RPS0B, 135-165 (162)	40S ribosomal protein S0-B	12 ± 4	10 ± 2	14 ± 6	13 ± 3	41 ± 9	68 ± 8	81 ± 9
P02994	TEF1, 356-369 (361;368)	Elongation factor 1-alpha	14 ± 3	18 ± 7	13 ± 6	13 ± 6	40 ± 6	50 ± 11	59 ± 11
P29509	TRR1, 138-153 (142;145)	Thioredoxin reductase 1	34 ± 3	39 ± 0	42 ± 3	44 ± 2	73 ± 0	80 ± 3	86 ± 1
P15731	UBC4, 103-129 (108)	Ubiquitin-conjugating enzyme E2 4	20 ± 8	15 ± 4	22 ± 5	24 ± 3	49 ± 6	73 ± 6	88 ± 6
F	**remain reduced during the course of the experiment**								
P07258	CPA1, 8-27 (11)	Carbamoyl-phosphate synthase small chain	22 ± 3	27 ± 7	24 ± 4	24 ± 7	27 ± 9	34 ± 8	30 ± 4
P03965	CPA2, 673-694 (678)	Carbamoyl-phosphate synthase large chain	13 ± 4	23 ± 6	11 ± 5	25 ± 3	21 ± 6	28 ± 8	35 ± 7
P39976	DLD3, 408-426 (409)	D-lactate dehydrogenase 3	10 ± 5	13 ± 5	20 ± 3	30 ± 12	20 ± 3	25 ± 10	21 ± 7
P14020	DPM1, 168-175 (172)	Dolichol-phosphate mannosyltransferase	7 ± 2	11 ± 6	13 ± 5	15 ± 4	19 ± 4	23 ± 8	27 ± 10
P32324	EFT2, 441-465 (448)	Elongation factor 2	5 ± 1	6 ± 1	9 ± 2	15 ± 1	14 ± 8	10 ± 3	16 ± 3
P07149	FAS1, 1296-1317 (1308)	Fatty acid synthase subunit beta	16 ± 7	25 ± 5	23 ± 3	23 ± 2	29 ± 6	19 ± 3	25 ± 4
P18562	FUR1, 163-172 (170)	Uracil phosphoribosyltransferase	18 ± 6	16 ± 6	6 ± 2	7 ± 2	16 ± 3	18 ± 8	14 ± 2
P18562	FUR1, 195-216 (215)	Uracil phosphoribosyltransferase	14 ± 1	17 ± 2	9 ± 4	18 ± 3	16 ± 4	26 ± 4	27 ± 5
P11986	INO1, 19-29 (19)	Inositol-1-phosphate synthase	31 ± 4	28 ± 7	19 ± 5	25 ± 3	33 ± 3	33 ± 7	30 ± 6
P15180	KRS1, 206-229 (212)	Lysyl-tRNA synthetase	16 ± 6	21 ± 8	12 ± 6	15 ± 3	21 ± 1	30 ± 9	28 ± 6
P53312	LSC2, 365-377 (365)	Succinyl-CoA ligase subunit beta	10 ± 3	15 ± 0	14 ± 11	20 ± 5	12 ± 4	25 ± 10	23 ± 11
P49367	LYS4, 335-344 (340)	Homoaconitase	31 ± 2	24 ± 5	21 ± 3	29 ± 4	33 ± 5	28 ± 2	26 ± 2
P38999	LYS9, 333-342 (340)	Saccharopine dehydrogenase	17 ± 6	13 ± 7	18 ± 9	9 ± 4	14 ± 7	15 ± 3	15 ± 2
P32179	MET22, 339-355 (349)	3'(2'),5'-bisphosphate nucleotidase	11 ± 5	15 ± 3	29 ± 3	27 ± 7	25 ± 7	20 ± 5	29 ± 6
P05694	MET6, 650-667 (657)	Methionine synthase	7 ± 2	7 ± 3	14 ± 7	8 ± 3	16 ± 5	13 ± 3	12 ± 0
P25293	NAP1, 231-250 (249)	Nucleosome assembly protein	10 ± 4	9 ± 3	8 ± 2	11 ± 2	24 ± 8	22 ± 6	21 ± 8
P38325	OM14, 28-35 (29)	Mitochondrial outer membrane protein	5 ± 1	13 ± 1	15 ± 3	14 ± 2	13 ± 5	12 ± 4	17 ± 4
P18239	PET9, 287-294 (288)	ADP,ATP carrier protein 2	11 ± 2	12 ± 8	11 ± 2	11 ± 2	14 ± 3	27 ± 7	20 ± 5
Q12189	RKI1, 73-83 (75)	Ribose-5-phosphate isomerase	13 ± 3	16 ± 5	17 ± 2	12 ± 5	17 ± 5	20 ± 4	24 ± 3
P04451	RPL23B, 13-32 (25)	60S ribosomal protein L23	13 ± 2	14 ± 8	20 ± 8	11 ± 4	22 ± 8	15 ± 6	21 ± 4
P26321	RPL5, 55-85 (62)	60S ribosomal protein L5	11 ± 5	26 ± 12	26 ± 12	17 ± 3	26 ± 1	29 ± 2	35 ± 6
P38886	RPN10, 110-127 (115)	26S proteasome regulatory subunit RPN10	15 ± 2	15 ± 1	16 ± 4	19 ± 6	18 ± 6	15 ± 1	16 ± 6
P10591	SSA1, 4-34 (15)	Heat shock protein SSA1	8 ± 1	8 ± 1	12 ± 2	13 ± 3	22 ± 8	15 ± 7	20 ± 4
P11484	SSB1, 431-455 (435;454)	Heat shock protein SSB1	12 ± 4	15 ± 7	10 ± 3	20 ± 6	32 ± 2	33 ± 6	29 ± 7
P32589	SSE1, 223-235 (228)	Heat shock protein homolog SSE1	7 ± 3	14 ± 5	16 ± 4	12 ± 4	11 ± 2	17 ± 7	24 ± 9
P53616	SUN4, 233-246 (239)	Septation protein SUN4	8 ± 3	11 ± 8	11 ± 5	13 ± 4	16 ± 4	14 ± 5	14 ± 3
P32861	UGP1, 351-369 (364)	UTP-glucose-1-phosphate uridylyltransferase	12 ± 5	25 ± 10	14 ± 3	16 ± 6	26 ± 3	21 ± 7	30 ± 8
P32527	ZUO1, 162-178 (167)	Zuotin	12 ± 2	14 ± 1	16 ± 2	18 ± 2	16 ± 2	24 ± 9	21 ± 6
G	**remain oxidized during the course of the experiment**								
O13547	CCW14, 38-61 (42;51;53)	Covalently-linked cell wall protein 14	94 ± 4	91 ± 1	89 ± 1	92 ± 2	92 ± 3	97 ± 2	92 ± 4
P00445	SOD1, 45-69 (58)	Superoxide dismutase [Cu-Zn]	55 ± 8	63 ± 6	56 ± 8	74 ± 3	83 ± 4	77 ± 11	83 ± 9
P14796	RPL40A, 13-21 (15;20)	60S ribosomal protein L40	62 ± 5	54 ± 12	62 ± 9	58 ± 12	75 ± 5	87 ± 5	94 ± 2
P15565	TRM1, 333-343 (333)	N(2),N(2)-dimethylguanos. tRNA methyltransferase	53 ± 2	71 ± 8	67 ± 12	72 ± 5	67 ± 14	67 ± 12	72 ± 9
P20081	FPR1, 50-64 (55)	FK506-binding protein 1	71 ± 5	70 ± 5	73 ± 8	82 ± 11	79 ± 8	81 ± 2	77 ± 14
P25373	GRX1, 25-45 (27;30)	Glutaredoxin 1	73 ± 4	81 ± 2	83 ± 3	76 ± 8	83 ± 10	86 ± 0	76 ± 0
P25491	YDJ1, 140-149 (143;146)	Mitochondrial import protein MAS5	56 ± 9	59 ± 5	70 ± 14	62 ± 4	85 ± 1	88 ± 3	95 ± 2
P32472	FPR2, 30-39 (36)	FK506-binding protein 2	80 ± 3	71 ± 3	72 ± 4	81 ± 7	81 ± 11	83 ± 6	86 ± 4

P41058	RPS29B, 33-40 (39)	40S ribosomal protein S29-B	51 ± 7	51 ± 10	64 ± 2	62 ± 10	81 ± 12	83 ± 7	95 ± 2
P53849	GIS2, 67-78 (67;70)	Zinc finger protein GIS2	51 ± 13	51 ± 2	64 ± 6	66 ± 12	72 ± 12	74 ± 3	73 ± 11
P53849	GIS2, 118-129 (118;121)	Zinc finger protein GIS2	59 ± 5	54 ± 14	71 ± 4	63 ± 0	72 ± 4	68 ± 10	74 ± 15
P53849	GIS2, 48-59 (49;52)	Zinc finger protein GIS2	71 ± 9	78 ± 9	64 ± 10	79 ± 6	78 ± 3	79 ± 4	80 ± 5
P87108	TIM10, 44-64 (44;61)	Mit. import inner memb. translocase subunit TIM10	90 ± 5	78 ± 10	83 ± 6	88 ± 4	93 ± 2	92 ± 6	94 ± 3
Q12512	ZPS1, 123-142 (123;130)	Protein ZPS1	91 ± 2	89 ± 3	81 ± 5	87 ± 3	85 ± 4	82 ± 8	90 ± 6
P50273	ATP22, 664-677 (672)	Mitochondrial translation factor ATP22	76 ± 10	68 ± 7	88 ± 3	76 ± 8	70 ± 8	73 ± 7	71 ± 3
H	minor increase in oxidation over several days								
P00330	ADH1, 277-299 (277;278)	Alcohol dehydrogenase 1	14 ± 4	23 ± 8	30 ± 1	27 ± 2	44 ± 7	48 ± 8	54 ± 5
P54839	ERG13, 289-301 (300)	Hydroxymethylglutaryl-CoA synthase	16 ± 5	18 ± 5	30 ± 4	44 ± 8	55 ± 8	70 ± 10	84 ± 11
P14540	FBA1, 96-115 (112)	Fructose-bisphosphate aldolase	9 ± 3	9 ± 5	18 ± 3	13 ± 7	29 ± 8	39 ± 2	41 ± 12
Q12154	GET3, 314-322 (317)	ATPase GET3	15 ± 4	16 ± 7	16 ± 4	19 ± 9	18 ± 9	46 ± 4	40 ± 14
P00498	HIS1, 149-179 (160)	ATP phosphoribosyltransferase	9 ± 3	7 ± 2	16 ± 5	14 ± 6	31 ± 6	41 ± 10	53 ± 7
P19882	HSP60, 266-287 (283)	Heat shock protein 60	15 ± 2	22 ± 4	30 ± 12	22 ± 4	23 ± 6	48 ± 7	54 ± 6
P33416	HSP78, 325-337 (333)	Heat shock protein 78	14 ± 4	15 ± 2	17 ± 5	30 ± 4	49 ± 8	49 ± 4	81 ± 15
P27809	KRE2, 123-130 (124)	Glycolipid 2-alpha-mannosyltransferase	29 ± 2	21 ± 8	30 ± 5	48 ± 2	54 ± 9	84 ± 5	88 ± 10
P15180	KRS1, 481-496 (486)	Lysyl-tRNA synthetase	14 ± 7	16 ± 4	14 ± 5	25 ± 12	19 ± 3	39 ± 6	47 ± 9
Q12122	LYS21, 194-206 (202)	Homocitrate synthase	16 ± 2	16 ± 2	8 ± 3	36 ± 8	26 ± 7	38 ± 4	49 ± 5
Q12122	LYS21, 289-319 (314)	Homocitrate synthase	14 ± 2	28 ± 10	17 ± 1	27 ± 9	31 ± 13	47 ± 11	53 ± 3
P05694	MET6, 563-573 (571)	Methionine synthase	16 ± 3	31 ± 4	32 ± 6	18 ± 4	33 ± 9	31 ± 4	51 ± 8
P06169	PDC1, 213-224 (221;222)	Pyruvate decarboxylase isozyme 1	15 ± 4	15 ± 3	20 ± 7	15 ± 8	32 ± 7	43 ± 8	43 ± 10
Q12672	RPL21B, 101-108 (101)	60S ribosomal protein L21-B	10 ± 3	28 ± 11	14 ± 2	26 ± 2	32 ± 4	34 ± 6	42 ± 3
P32827	RPS23B, 83-109 (92)	40S ribosomal protein S23	11 ± 2	5 ± 2	8 ± 1	16 ± 6	29 ± 5	30 ± 10	39 ± 5
P26783	RPS5, 85-92 (87)	40S ribosomal protein S5	36 ± 1	33 ± 5	44 ± 5	55 ± 10	69 ± 5	52 ± 8	61 ± 6
P38788	SSZ1, 65-85 (81)	Ribosome-associated complex subunit SSZ1	17 ± 6	21 ± 1	28 ± 3	26 ± 3	17 ± 1	31 ± 7	49 ± 6
P04801	THS1, 264-278 (268)	Threonyl-tRNA synthetase	30 ± 5	27 ± 5	17 ± 2	30 ± 2	48 ± 3	58 ± 9	70 ± 12
P25491	YDJ1, 181-200 (185;188)	Mitochondrial protein import protein MAS5	45 ± 3	44 ± 3	51 ± 7	70 ± 3	73 ± 16	83 ± 4	88 ± 1

Table 4.C. Side by side comparison of the thiol oxidation status of proteins from chronological aging yeast cultivated in different glucose conditions. Data from Appendix, Table 4.A. and 4.B. is combined with the thiol oxidation status of Day 3, 5, and 10 of yeast cells cultivated in 2% glucose SCD media for 2 days and then shifted into water (last three columns). For side by side comparison, peptides in SCD media are grouped by their gene designation and cluster, while peptides identified in CR media and peptides from cells shifted into water are grouped in the order of the peptides in SCD media. For protein names and Swiss-Prot accession numbers, refer to Appendix, Table 4.A. and 4.B. "% oxidation" denotes the average oxidation of the redox-sensitive cysteine (first column, in brackets), obtained from three independent experiments ± standard deviation. *: peptide identical in both isoenzymes; a,b,c.: at this mass, two different peptides were identified; §: peptides is grouped into same cluster in 2% glucose SCD media and 0.5% glucose CR media. Missing values are due to insufficient MS-quantification.

Gene, Peptide (Cys)	cluster	2% glucose SCD media % oxidation at...					cluster	0.5% glucose CR media % oxidation at...							SCD → water % oxidation at...		
		log phase	Day 1	Day 2	Day 3	Day 4		log phase	Day 1	Day 2	Day 3	Day 4	Day 5	Day 7	Day 3	Day 5	Day 10
ACO1, 376-392 (382)[§]	A	7	4	12	69	84	A	7	11	11	20	28	84	88	26 ± 5	27 ± 5	26 ± 1
ADE17, 285-304 (300)[§]	A	8	10	9	79	75	A	7	9	16	18	28	66	79			
ADH6, 141-174 (163)	A	24	23	31	4505	66											
ADO1, 121-130 (121)[§]	A	11	14	12	88	92	E	12	13	9	13	39	79	80			
AHP1, 114-124 (120)[§]	A	6	5	15	73	89	A	11	6	10	13	25	76	86			
AHP1, 49-79 (62)[§]	A	12	17	18	60	80	A	16	15	17	17	22	71	93			
ALD6, 124-134 (132)[§]	A	15	14	18	66	89	A	13	13	13	16	31	83	88			
ALD6, 290-311 (306)[§]	A	15	15	23	51	67	A	25	39	21	22	26	60	88	31 ± 1	26 ± 5	28 ± 4
ALO1, 359-374 (359)[§]	A	22	33	2	81	93	A	20	25	34	22	34	68	84			
APA1, 127-139 (129)	A	7	6	6	53	87	B	8	15	7	12	18	9	70			
ARO4, 240-249 (244)[§]	A	13	17	23	56	83	A	18	12	23	17	18	61	79	29 ± 7	27 ± 3	32 ± 3
ASN1, 421-440 (432)[§]	A	10	14	7	67	76	A	7	11	15	28	30	52	60			
ASN1/2, 396-406 (404)[§,*]	A	7	9	15	78	83	A	12	12	11	14	24	63	66			
BAT1, 125-141 (126)[§]	A	11	9	15	72	87	A	10	9	9	15	26	76	90			
BAT1, 220-237 (227)[§]	A	8	7	22	71	90	A	8	10	17	14	30	60	85			
BMH1, 77-88 (85)[§]	A	16	17	25	84	92	A	10	11	14	17	32	96	90	26 ± 3	18 ± 7	21 ± 4
CCT6, 515-535 (530)[§]	A	5	4	13	63	82	A	11	4	4	17	27	63	66			
CCT6, 79-103 (96)	A	10	12	13	62	80	A	9	6	9	14	27	71	94			
CDC19, 287-309 (296)	A	12	15	22	73	88											
CDC19, 370-394 (371)[§]	A	18	22	25	84	87	A	28	10	13	24	30	90	86			
CHA1, 175-195 (178)[§]	A	11	11	15	53	75	E	9	9	12	15	43	70	81			
CPA2, 159-182 (170)[§]	A	10	7	12	55	79	A	12	14	13	14	27	78	81	19 ± 2	19 ± 6	26 ± 5
CPA2, 673-694 (678)	A	18	21	20	81	83	F	13	23	11	25	21	28	35	34 ± 5	23 ± 1	26 ± 2

Chapter 4: Collapse of the cellular redox balance in aging yeast

		7	3	6	72	93		51	34	43	20	27	87	93			
CPR6, 150-165 (156)	A	7	33	22	44	72	A	93	51	34	43	20	27	87	45 ± 1	48 ± 4	46 ± 9
CPS1, 361-370 (368)	A	33	22	44	77	90	C	40	40	46	37	43	62	74	28 ± 2	17 ± 2	26 ± 3
CRP1, 105-118 (113)	A	21	16	33	67	79	A	25	16	20	20	39	83	85			
DBP2, 215-226 (219)	A	15	22	30	74	73											
DED1, 256-270 (257)	A	11	11	18	83	81	A	15	16	9	21	27	70	95	31 ± 3	26 ± 6	26 ± 4
DPS1, 250-259 (255)	A	28	22	31	76	86	A	29	29	31	31	35	70	81	29 ± 4	26 ± 3	28 ± 5
DUS3, 210-225 (224)	A	13	19	18	74	98	A	9	14	15	23	34	70	89			
EFT2, 353-370 (366)	A	12	6	13	78	88	A	12	5	13	14	25	73	81	16 ± 6	11 ± 2	15 ± 3
EFT2, 121-144 (136)	A	11	10	26	70	86	A	13	17	17	17	30	77	83			
EFT2, 724-749 (735)	A	8	6	15	77	85	A	9	8	13	20	19	70	78			
ENO2, 244-255 (248)	A	5	3	10	51	87	A	8	9	10	11	19	55	82	27 ± 2	26 ± 2	24 ± 4
ENP1, 320-333 (330)	A	11	4	12	67	96	A	9	11	10	16	22	56	79			
ERG10, 343-362 (358)	A	12	12	19	56	82	A	7	6	8	15	30	76	83			
ERG11, 137-146 (142)	A	15	13	8	62	81	A	18	15	20	17	29	69	60	35 ± 4	39 ± 4	40 ± 5
ERG9, 340-350 (341)	A	9	12	24	72	62	A	9	12	11	21	26	69	80			
FAS2, 900-919 (917)	A	22	16	30	62	86	D	14	8	15	59	81	87	72			
FBA1, 209-230 (226)	A	12	8	22	64	79											
FBA1, 250-264 (254)	A	13	16	22	63	76	A	11	11	15	24	44	74	81			
FRDS1, 446-459 (454)	A	6	11	25	60	90	A	7	16	15	27	32	66	85	25 ± 6	13 ± 6	24 ± 8
FRS1, 121-138 (137)	A	12	15	17	74	86	A	9	23	11	15	30	67	75			
GCD1, 163-188 (172)	A	13	18	16	85	79	A	14	13	15	26	29	72	88			
GCN1, 1070-1078 (1077)	A	22	26	33	58	71	E	15	16	18	24	48	47	63			
GDI1, 217-226 (221)	A	14	22	20	65	89	A	10	13	19	29	40	61	81	22 ± 3	28 ± 4	25 ± 3
GET3, 314-322 (317)	A	15	15	20	50	95	H	15	16	16	19	18	46	40			
GFD2, 88-103 (90)	A	16	13	22	58	85	A	16	14	11	16	32	96	92	25 ± 4	27 ± 6	24 ± 5
GLC7, 20-40 (38)	A	9	9	24	77	74	B	8	7	12	16	30	73	88			
GND1, 287-295 (287)	A	15	9	15	54	71	A	15	14	18	24	24	39	56	14 ± 6	20 ± 3	15 ± 3
GPP1/GPP2, 210-222 (211)	A	27	22	38	59	69	A	18	24	30	27	51	75	91	44 ± 4	42 ± 7	40 ± 2
HEM15, 118-136 (123)	A	27	28	39	61	77	A	21	13	19	24	33	62	83			
HOM2, 235-249 (247)	A	13	11	20	68	86	A	11	9	15	14	27	54	86	26 ± 4	27 ± 5	36 ± 8
HOM2, 250-263 (259)	A	20	15	19	64	76	A	19	13	17	24	29	57	73	19 ± 3	18 ± 3	22 ± 3
HOM3, 38-56 (53)	A	5	6	18	67	80	A	6	16	23	18	33	57	79	27 ± 3	25 ± 3	32 ± 5
IDP1, 360-371 (370)	A	10	3	9	69	84	B	17	12	24	29	13	34	57			
IDP1, 89-98 (89)	A	13	9	32	65	76	A	19	17	19	25	44	70	84			
ILS1, 99-120 (119)	A	25	22	21	95	89											
ILV3, 157-180 (176)	A	14	22	27	61	73	A	22	12	18	27	32	77	83	18 ± 4	14 ± 3	16 ± 3
ILV5, 307-321 (316)	A	15	12	17	51	82	A	18	9	20	14	26	90	87			
ILV6, 102-120 (111)	A	8	17	18	65	81	A	9	8	7	14	25	85	86	24 ± 3	27 ± 3	31 ± 2
ILV6, 261-272 (270)	A	9	8	17	70	84	A	8	10	13	13	23	65	73			
KAP123, 977-987 (984)	A	7	5	18	69	98	A	7	7	9	11	18	61	84	17 ± 2	24 ± 3	24 ± 0
KRE2, 123-130 (124)	A	14	17	27	67	82	H	19	18	20	23	17	55	83	40 ± 2	30 ± 3	29 ± 1
KRE2, 123-130 (124)	A	33	26	49	80	97	A	29	21	30	48	54	84	88			
KRS1, 429-446 (438)	A	9	4	18	63	78	A	9	17	11	15	27	66	81	29 ± 4	18 ± 4	27 ± 11
KRS1, 430-446 (438)	A	10	12	22	75	86	A	9	26	20	28	36	79	83			

Protein																		
KRS1, 481-496 (486)	A	13	8	16	58	68		14	16	14	25	19	39	47	17 ± 3	27 ± 5	25 ± 2
LEU1, 230-242 (233)	A	15	22	29	67	80	H	13	9	14	16	15	27	64	31 ± 6	33 ± 1	33 ± 2
LEU1, 60-77 (63)	A	22	20	24	62	85	B	13	15	23	22	42	87	85	21 ± 1	18 ± 6	20 ± 2
LEU4, 345-353 (345)	A	10	14	18	67	79	A	8	9	11	14	26	65	80			
LPD1, 45-58 (54)	A	12	17	24	72	80	A	15	18	18	29	35	73	88			
LSC2, 365-377 (365)	A	12	10	13	68	81	F	10	15	14	20	12	25	23			
MAE1, 535-555 (541)	A	10	9	19	67	86	A	10	15	19	19	17	71	90			
MCM6, 813-832 (815)	A	20	15	16	50	75	A	5	18	13	16	27	71	88			
MET6, 563-573 (571)	A	14	9	26	64	89	H	16	31	32	18	33	31	51			
MRPL19, 36-46 (40)	A	7	7	6	60	87	A	9	9	8	14	25	75	77			
NIF3, 105-123 (114)	A	10	7	16	68	64	A	12	9	14	16	24	51	69			
NOP1, 273-288 (275)	A	9	15	12	73	92	A	13	14	11	16	29	69	83			
NOP1, 312-319 (314)	A	9	9	9	69	83	A	10	14	9	21	36	71	81			
NSR1, 233-254 (241)	A	14	31	17	73	77	A	12	22	30	32	34	67	79			
NUP116, 1030-1042 (1031)	A	9	17	21	52	70											
OLA1, 126-143 (126)	A	5	6	13	81	76	A	6	8	6	13	25	66	81	25 ± 7	31 ± 6	31 ± 2
OLA1, 72-81 (76)	A	9	9	13	60	83	A	6	16	15	20	26	65	85	15 ± 4	12 ± 1	28 ± 5
PDC1, 213-224 (221;222)	A	22	18	13	58	61	H	15	15	20	15	32	43	43			
PDC5, 128-154 (152)	A	11	15	12	77	78	A	14	10	14	16	30	77	88			
PET9, 66-76 (73)	A	7	6	10	51	80	A	11	4	12	17	33	76	66			
PET9, 265-273 (271)	A	16	14	18	75	89	A	8	10	11	14	26	57	81	22 ± 2	23 ± 6	17 ± 4
PET9, 67-76 (73)	A	13	15	12	77	71	A	9	10	11	14	29	78	86	19 ± 6	26 ± 4	25 ± 2
PFY1, 77-91 (89)	A	7	5	10	72	92	A	5	7	14	13	24	73	94			
PGK1, 83-108 (98)	A	9	5	8	62	86	A	11	8	10	12	23	75	88	15 ± 3	24 ± 3	18 ± 3
PGM2, 372-389 (376)	A	12	9	13	87	82	A	10	7	12	10	11	63	85			
PMA1/2, 392-414 (376;405)	A	13	12	18	72	62	A	13	12	10	10	19	54	73			
PMA1/2, 415-443 (409;438)	A	12	16	16	68	85	A	10	6	12	11	19	59	83			
PMA1/2, 487-503 (472;501)	A	18	17	23	69	97	A	17	14	11	14	26	87	81	37 ± 3	36 ± 3	30 ± 4
POR1, 125-132 (130)	A	8	11	14	77	86	A	7	7	11	14	28	55	70			
PPT1, 244-262 (247)	A	12	15	27	52	70											
PRE10, 73-86 (76)	A	12	13	17	69	73	A	17	15	19	22	29	77	80	21 ± 4	9 ± 3	15 ± 5
PSA1, 100-125 (113)	A	12	4	11	56	77											
QCR2, 129-152 (137)	A	23	32	35	84	90	A	20	24	15	18	35	88	81			
RAD23, 364-371 (365)	A	23	30	26	50	82	A	17	25	16	27	35	70	90	27 ± 6	15 ± 7	22 ± 4
RNA1, 169-182 (180)	A	14	16	27	78	90	A	9	15	26	22	36	82	81			
RNR4, 221-235 (228)	A	12	19	19	67	81	E	12	16	13	15	22	68	92			
RPL10, 41-69 (49)	A	15	15	31	73	89	A	21	17	28	23	59	80	90	26 ± 7	18 ± 4	24 ± 2
RPL10, 70-82 (71)	A	21	16	23	71	86	A	18	13	16	26	31	60	86	21 ± 3	31 ± 1	40 ± 8
RPL12B, 131-142 (141)	A	8	7	15	57	75	A	10	20	21	15	26	63	78			
RPL21B, 101-108 (101)	A	9	9	21	62	88	H	5	28	14	26	32	34	42			
RPL40A, 39-48 (39)	A	18	21	37	73	90	A	18	19	31	20	38	69	81	33 ± 4	24 ± 2	27 ± 4
RPS0B, 135-165 (162)	A	18	17	23	70	87	E	12	10	14	13	41	68	81	31 ± 4	34 ± 6	31 ± 3
RPS11B, 117-133 (128)	A	18	15	24	78	77	A	15	17	15	24	19	77	74			
RPS14B, 72-83 (72)	A	12	14	19	73	88	A	10	18	16	22	32	56	89			

Chapter 4: Collapse of the cellular redox balance in aging yeast   173

Peptide																		
RPS17B, 34-45 (35)§	A	8	5	6	61	89	A	8	6	12	15	25	84	88				
RPS1B, 65-83 (69)§	A	8	5	6	60	76	A	5	7	13	15	28	68	89				
RPS21B, 16-22 (17)	A	20	15	16	56	65	B	16	14	16	21	22	47	57				
RPS37, 68-73 (68)§	A	24	18	25	55	76	A	20	21	35	25	23	66	77	27 ± 6	21 ± 3	25 ± 3
RPS8B, 165-178 (168)§	A	9	15	18	68	84	A	10	16	19	25	30	62	74	22 ± 2	18 ± 2	22 ± 1
RVB2, 221-232 (224)	A	13	19	27	63	74	B	15	22	13	17	30	39	64			
SAH1, 192-199 (198)§	A	16	33	28	50	84	A	19	22	19	21	40	75	70			
SAH1, 218-236 (231)§	A	12	27	18	86	86	A	8	16	17	26	33	76	90	19 ± 3	26 ± 4	25 ± 6
SCL1, 63-77 (74)§	A	4	20	14	71	88	A	14	23	13	15	28	60	85	22 ± 1	22 ± 1	25 ± 1
SDH1b, 613-632 (624)§	A	21	15	15	72	83	A	13	12	19	19	36	87	85			
SER1, 182-202 (182)§	A	11	9	23	55	86	A	8	8	13	14	25	60	72			
SES1, 363-379 (370,373)§	A	26	30	44	69	70	A	21	29	36	32	42	58	74	39 ± 6	26 ± 5	35 ± 6
SHM1, 89-106 (103)§	A	14	12	19	50	83	A	13	11	15	14	27	79	88	15 ± 4	20 ± 5	17 ± 1
SHM2, 188-198 (197)	A	23	15	28	62	87	B	15	19	33	18	24	44	78			
SHP1, 78-103 (88)§	A	13	9	18	51	79	A	14	16	14	17	33	66	78	22 ± 2	22 ± 1	26 ± 2
SSA2, 14-34 (15)	A	17	17	13	73	99	B	13	16	13	22	20	47	81			
SSA2, 304-313 (308)§	A	10	19	16	54	94	A	14	10	11	11	34	83	83	21 ± 6	14 ± 6	22 ± 7
SSB1, 431-451 (435)§	A	26	22	28	70	81	A	24	21	22	26	28	72	87			
SSB2, 452-466 (454)§	A	18	14	23	62	80	A	15	19	15	14	29	81	88	19 ± 6	20 ± 5	28 ± 1
SSE2, 375-388 (380)§	A	11	8	22	85	86	A	13	14	16	21	22	67	84			
SUB2, 359-368 (360)§	A	11	17	12	50	68	B	17	16	20	15	31	49	73			
TDH2/3, 144-160 (150;154)¥,§	A	7	4	10	80	87	A	12	14	9	17	20	80	81	37 ± 5	34 ± 4	41 ± 7
TEF1, 21-37 (31)§	A	6	7	5	74	88	A	10	6	18	19	30	73	89	26 ± 6	11 ± 2	16 ± 4
TEF1, 320-328 (324)§	A	6	5	7	60	98	A	8	12	18	16	36	79	86	14 ± 5	8 ± 3	10 ± 2
TEF3, 665-696 (691)	A	21	9	13	53	82	A	12	8	14	23	22	66	70			
TEF3, 708-732 (731)§	A	26	30	32	52	60	H	30	27	17	30	48	58	70	35 ± 1	24 ± 1	33 ± 6
TFP1, 356-367 (358)§	A	12	4	14	63	86	A	13	9	11	14	22	82	86	28 ± 2	18 ± 2	25 ± 4
THR1, 216-231 (216)§	A	10	13	16	63	78	A	8	18	15	27	29	67	95			
THR4, 416-432 (428)§	A	17	18	32	61	75	A	16	19	30	16	32	57	78			
THS1, 264-278 (268)	A	7	9	10	56	63	B	4	5	5	10	10	30	78	28 ± 7	32 ± 4	25 ± 3
TIF11, 83-101 (89)	A	7	4	5	68	93	A	5	3	8	14	27	74	90			
TIM44, 356-370 (369)§	A	6	7	10	50	76	B	12	30	13	19	21	45	76			
TOM40, 324-330 (326)§	A	6	5	7	63	87	A	7	7	14	16	25	63	86	22 ± 7	19 ± 7	25 ± 8
TPI1, 115-134 (126)	A	7	13	21	96	86	A	15	23	30	29	25	79	91	22 ± 5	22 ± 5	29 ± 1
TPI1, 27-55 (41)§	A	27	22	29	55	84	A	14	17	27	15	31	20	54			
TPS1, 206-222 (210)§	A	11	9	24	71	80	A	12	25	14	16	26	21	30			
TRP2, 121-135 (133)§	A	11	17	13	73	99	B	12	11	9	20	37	60	89			
TRP5, 402-422 (420)§	A	5	13	10	74	89	A	6	7	13	21	26	71	60	22 ± 4	33 ± 4	24 ± 1
UBA1, 596-603 (600)	A	23	20	32	64	81	A	14	17	19	14	27	62	83			
UGP1, 351-369 (364)§	A	17	14	21	73	82	A	14	12	14	15	30	66	79	22 ± 7	16 ± 3	23 ± 1
URA2, 1832-1854 (1832)§																	
URA5, 11-23 (19)§																	
VAC8, 341-359 (352)§																	
VMA2, 170-189 (188)§																	

174 Chapter 4: Collapse of the cellular redox balance in aging yeast

Peptide																		
VPH1, 386-403 (394,396)§	A	11	8	22	62	89	A	12	10	21	18	23	66	67	26±4	21±2	24±0
YDJ1, 364-382 (370)§	A	7	10	9	68	82	A	8	16	11	13	25	75	77	12±5	15±2	26±3
YDR539W, 58-74 (69)§	A	15	19	22	74	90	A	15	21	27	31	51	63	88			
YGL157W, 232-247 (240)	A	31	19	30	70	91											
YHR020W, 387-405 (396)§	A	9	4	18	61	80	A	13	19	7	15	25	70	81			
YLR179C, 137-146 (142)	A	4	3	13	75	91											
YOR285W, 93-103 (98)§	A	21	9	18	67	79	A	18	17	13	17	32	66	74			
YRB1, 134-154 (135)§	A	16	9	17	69	85	A	16	12	16	14	27	60	76			
ACC1, 1212-1241 (1225)§	B	16	4	19	44	66	B	10	20	19	19	14	25	61			
ARC1, 261-274 (266)§	B	17	14	24	45	70	A	11	7	13	17	34	77	81			
CCT3, 512-531 (518)§	B	15	12	18	35	39	A	15	20	27	31	51	40	88			
CCT7, 452-469 (454)§	B	11	31	16	27	59	B	7	20	13	19	32	39	80			
CDC48, 76-93 (92)§	B	15	14	23	42	59	B	13	12	18	24	36	71	77			
CPR1, 36-53 (38)^4,§	B	12	14	19	44	63	A	9	6	13	17	34	10	16			
EFT2, 441-465 (448)	B	8	14	17	31	33	F	5	6	9	15	14	26	55			
FAA4, 26-52 (49)§	B	4	24	13	33	83	B	11	15	13	22	30	33	68			
FBA1, 96-115 (112)	B	9	4	14	34	76	B	8	10	10	15	19	39	41			
GUA1, 438-455 (450)§	B	13	5	14	30	39	H	9	9	18	13	29	42	77			
GUS1, 378-400 (385)§	B	9	18	12	30	67	B	8	12	14	15	27	31	56			
HIS1, 149-179 (160)	B	15	11	25	43	64	B	11	14	14	17	23	41	53			
HSP104, 387-407 (400)§	B	9	12	18	46	69	H	9	7	16	14	31	36	87			
HSP78, 325-337 (333)	B	13	7	12	28	57	B	11	5	9	14	32	49	81	12±6	13±4	20±3
HXK2, 395-407 (398,404)	B	17	9	15	33	65	H	14	15	17	30	49	68	89	21±3	10±1	16±2
ILV6, 236-254 (244)§	B	15	16	25	44	57	A	17	15	22	30	38	41	69			
KRS1, 206-229 (212)	B	12	9	22	40	55	B	7	14	16	23	25	30	28			
LYS21, 194-206 (202)	B	13	8	17	24	37	F	16	21	12	15	21	38	49	23±6	17±5	18±1
LYS21, 269-319 (314)	B	14	13	21	32	43	H	16	16	8	36	26	47	53			
MIS1, 216-229 (224)§	B	10	20	24	46	61	H	14	28	17	27	31	44	60			
PUP2, 94-122 (117)§	B	23	14	19	35	53	B	5	11	12	16	32	47	79			
RKI1, 73-83 (75)	B	12	15	18	42	67	B	20	19	17	27	30	20	24			
RPL11B, 144-153 (144)§	B	8	11	14	16	49	F	13	16	12	12	17	41	62			
RPL35B, 63-77 (67)	B	8	11	11	23	54	D	14	14	8	18	24	75	93	22±4	12±4	26±0
RPL5, 55-85 (62)§	B	13	27	17	27	60	F	16	6	17	17	48	29	35			
RPS23B, 83-109 (92)	B	8	10	15	29	75	H	11	26	26	26	26	30	39			
RPS8B, 179-195 (179)§	B	14	11	14	29	34	B	5	5	8	16	29	44	71	19±7	20±6	17±7
SEC65, 98-110 (104)	B	10	10	12	38	50	A	11	11	13	17	30	80	77			
SES1, 397-410 (400)	B	11	14	17	37	70	A	15	13	8	16	31	67	77			
SIS2, 2-23 (21)	B	18	24	21	42	53	A	12	12	12	10	25	33	29			
SSB1, 431-455 (435,454)§	B	19	18	26	43	38	F	12	15	10	20	32	17	24			
SSE1, 223-235 (228)	B	10	16	27	42	45	F	7	14	16	12	11	40	87			
SSH1, 70-77 (74)§	B	7	10	18	32	82	B	6	7	13	15	27					

Chapter 4: Collapse of the cellular redox balance in aging yeast

Protein		7	14	8	43	62		9	8	12	14	26	48	61			
TEF4, 125-143 (129)§	B						B							61	21 ± 2	15 ± 5	18 ± 3
VAC8, 428-447 (445)§	B	12	10	12	42	54	B	11	11	7	17	34	45	73			
CDC19, 414-425 (418)§	C	46	39	49	76	86	C	41	42	35	36	46	72	83	57 ± 1	43 ± 6	52 ± 2
GIS2, 67-78 (67,70)	C	52	39	57	79	83	G	51	51	64	66	72	74	73	48 ± 5	44 ± 4	48 ± 7
RIB3, 125-136 (133)	C	45	36	53	85	97	E	37	29	37	37	59	61	86			
RPL37B, 33-43 (34-37)	C	40	36	40	66	75	E	38	34	29	39	65	78	88			
RPS26B, 67-82 (74,77)§	C	46	54	55	62	87	C	51	50	64	59	64	63	78	49 ± 4	57 ± 5	58 ± 4
RPS29A, 23-32 (24)§	C	47	39	53	83	90	C	49	42	58	57	60	78	85			
RPS29A, 33-40 (39)§	C	46	56	59	69	88	C	45	47	55	59	71	88	88	66 ± 3	67 ± 6	68 ± 4
RPS29B, 33-40 (39)	C	55	47	63	87	92	G	51	51	64	62	81	83	95	54 ± 2	61 ± 5	45 ± 5
RPS31, 44-62 (45,50)§	C	46	42	58	83	78	C	41	45	36	50	57	64	84	39 ± 2	42 ± 7	50 ± 2
RPS5, 85-92 (87)	C	40	42	54	68	73	H	36	33	44	55	69	52	61			
SSO2, 108-123 (122)§	C	41	34	46	59	67	C	44	40	28	32	46	59	74			
CCT4, 393-402 (399)	D	12	24	58	87	86	E	12	14	13	11	39	72	82	42 ± 5	36 ± 7	32 ± 3
CDC48, 104-119 (115)	D	12	29	55	75	93	E	13	15	14	28	62	68	90			
ERG13, 289-301 (300)	D	17	38	47	80	91	H	16	18	30	44	55	70	84			
FUS2, 366-375 (371)	D	18	33	57	68	80	A	9	18	12	20	31	64	91			
LAP4, 191-203 (202)	D	19	38	45	83	86	A	25	35	27	22	39	61	81	41 ± 6	35 ± 5	52 ± 7
PYC2, 211-222 (218)§	D	11	28	46	81	93	D	13	11	22	47	66	77	81	51 ± 7	46 ± 6	42 ± 5
RPL4B, 85-95 (94)	D	11	25	34	33	92											
TEF1, 409-421 (409)	D	13	38	40	62	70	A	18	30	17	26	36	57	68	39 ± 3	26 ± 6	27 ± 7
ARO2, 210-224 (221)	E	13	13	31	71	74											
CCT8, 335-343 (336)	E	47	40	77	76	89											
HEM1, 386-391 (386)	E	13	21	58	77	88											
IDP1, 393-403 (398)§	E	18	19	41	61	85	E	16	18	32	25	49	81	80	42 ± 1	39 ± 5	44 ± 6
KGD1, 982-996 (983)§	E	25	21	49	57	84	E	18	18	19	16	57	54	70			
LYS2, 611-625 (614)	E	15	22	49	79	86	D	14	11	12	48	56	74	81			
MES1, 353-366 (353)	E	23	34	64	86	82	D	20	21	31	61	78	80	98	50 ± 5	49 ± 2	52 ± 4
OLA1, 43-62 (43)§	E	17	12	46	91	83	E	20	10	22	29	53	78	88			
PRB1, 344-368 (361)§	E	18	23	57	91	94	E	12	14	26	23	77	80	88			
PUT2, 154-171 (162)	E	12	10	35	60	66											
RPL42B, 68-76 (74)§	E	15	22	49	86	86	D	15	19	25	49	54	77	93	44 ± 4	33 ± 2	36 ± 6
RPS11B, 58-67 (58)	E	18	17	37	77	81	A	14	17	27	25	36	83	86	38 ± 2	45 ± 6	38 ± 4
RPS22B, 72-78 (72)	E	13	9	34	62	72	A	32	29	36	32	41	72	79	29 ± 5	25 ± 4	28 ± 3
SES1, 411-433 (413,414)	E	22	23	55	66	79	D	20	19	17	43	65	71	81			
TRR1, 138-153 (142,145)§	E	33	33	65	77	82	E	34	39	42	44	73	80	86	43 ± 5	49 ± 6	48 ± 4
UBC4, 103-129 (108)§	E	19	23	48	70	81	E	20	15	22	24	49	73	88			
YCR090C, 112-125 (124)	E	34	39	67	76	88											
YDJ1, 181-200 (185,188)	E	46	46	82	89	75	H	45	44	51	70	73	83	88			

Gene						Group										
ATP3, 115-126 (117)	9	15	18	20	17	F	22	27	24	24	27	34	30	20 ± 3	24 ± 2	29 ± 2
CPA1, 8-27 (11)§	24	30	24	32	30	F	10	13	20	30	20	25	21	15 ± 5	19 ± 6	15 ± 1
DLD3, 408-426 (409)§	13	24	25	36	20	F	7	11	13	15	19	23	27			
DPM1, 168-175 (172)§	14	10	12	17	29	F	16	25	23	23	29	19	25			
FAS1, 1296-1317 (1308)§	13	18	23	21	25	F	14	17	9	18	16	26	27			
FUR1, 195-216 (215)§	11	11	15	16	21	F	31	28	19	25	33	33	30			
INO1, 19-29 (19)§	34	39	26	33	36	F	31	24	21	29	33	28	26	36 ± 4	36 ± 1	39 ± 1
LYS4, 335-344 (340)§	35	23	35	49	39	F	17	13	18	9	14	15	15			
LYS9, 333-342 (340)§	15	5	14	16	6	F	11	15	29	27	25	20	29	28 ± 7	23 ± 4	30 ± 2
MET22, 339-355 (349)§	21	14	13	23	22	F	11	7	14	8	16	13	12			
MET6, 650-667 (657)§	12	9	13	20	31	F	7									
MRPL4, 62-76 (62)	9	15	15	26	24	F	12	14	16	18	16	22	21	7 ± 1	21 ± 7	10 ± 4
NAP1, 231-250 (249)§	16	17	19	29	35	F	10	9	8	11	13	12	17			
OM14, 28-35 (29)§	4	5	28	17	15	F	5	13	15	14	14	27	20			
PET9, 287-294 (288)§	13	12	18	21	22	F	11	12	11	11	22	15	21			
RPL23B, 13-32 (25)§	15	24	18	16	11	F	13	14	20	11	18	15	16			
RPN10, 110-127 (115)§	17	18	13	14	23	F	15	15	16	19	22	15	20			
SSA1, 4-34 (15)§	13	13	17	25	21	F	8	8	12	13	16	14	14			
SUN4, 233-246 (239)§	5	11	5	9	9	F	8	11	11	16						
VAS1, 998-1031 (1015)§	15	15	22	26	32	F										
ZUO1, 162-178 (167)§	13	13	12	15	19	F	12	14	16	18	16	24	21	17 ± 1	19 ± 2	17 ± 3
ATP22, 664-677 (672)§	75	64	63	58	54	G	76	68	88	76	70	73	71			
CCW14, 38-61 (42;51;53)§	93	96	85	88	93	G	94	91	89	92	92	97	92			
FPR1, 50-64 (55)§	76	73	70	83	79	G	71	70	73	82	79	81	77			
FPR2, 30-39 (36)§	79	73	76	83	83	G	80	71	72	81	81	83	86			
GIS2, 118-129 (118;121)§	70	59	67	80	78	G	59	54	71	63	72	68	74			
GIS2, 48-59 (49;52)§	66	70	73	76	78	G	71	78	64	79	78	79	80			
GRX1, 25-45 (27;30)§	80	83	91	93	97	G	73	81	83	76	83	86	76			
KRE6, 476-493 (479;481)§	92	90	94	95	94	G										
RPL40A, 13-21 (15;20)§	66	63	79	88	79	G	62	54	62	58	75	87	94	73 ± 4	68 ± 3	62 ± 6
SOD1, 138-154 (147)§	72	63	72	80	95	C	52	47	56	60	58	83	93	64 ± 6	63 ± 2	69 ± 6
SOD1, 45-69 (58)§	69	83	84	86	95	G	55	63	56	74	83	77	83			
TIM10, 44-64 (44;61)§	93	99	97	95	95	G	90	78	83	88	93	92	94	64 ± 3	94 ± 1	92 ± 2
TRM1, 333-343 (333)§	65	68	64	63	83	G	53	71	67	72	67	67	72	63 ± 3	62 ± 4	58 ± 4
YDJ1, 140-149 (143;146)§	64	53	70	68	76	G	56	59	70	62	85	88	95			
ZPS1, 123-142 (123;130)§	96	91	91	98	94	G	91	89	81	87	85	82	90			
ADH1, 277-299 (277;278)§	17	32	42	45	39	H	14	23	30	27	44	48	54	25 ± 5	29 ± 5	24 ± 2
ERG1, 170-189 (174)§	27	21	46	55	73	B										
GLK1, 440-460 (448)§	15	13	28	42	62	H	11	10	11	14	27	36	70			

Peptide		5	13	31	29	40			15	22	30	22	23	48	54	40 ± 2	26 ± 4	33 ± 6
GRE2, 77-105 (86)	H							H								40 ± 2	26 ± 4	33 ± 6
HSP60, 266-287 (283) s	H	18	14	34	31	40		H	18	15	25	52	61	71	83	39 ± 5	47 ± 6	57 ± 9
MAE1, 202-221 (220)	H	29	20	43	43	69		C	40	33	51	43	63	67	88	51 ± 7	58 ± 3	59 ± 0
PAA1, 48-61 (51;55)	H	26	36	53	48	61		A	27	27	45	33	41	84	84			
RPL34B, 44-60 (44-47)	H	43	33	52	64	79												
RPL40A, 8-17 (15)	H	26	29	44	77	84												
RPL8B, 150-171 (170)	H	9	14	26	31	69		C	45	36	50	46	62	87	83	62 ± 6	57 ± 2	57 ± 4
RPS29B, 23-32 (24)	H	41	46	66	89	98		D	38	23	38	63	67	76	84			
SNQ2, 1097-1121 (1098)	H	30	62	42	45	86		H	17	21	28	26	17	31	49	24 ± 2	27 ± 5	30 ± 3
SSZ1, 65-85 (81)	H	29	27	42	58	50		A	17	15	15	17	23	53	67	27 ± 3	15 ± 7	18 ± 4
VAC8, 300-317 (315)	H	18	16	30	36	75												
DED1, 275-285 (276)								A	11	18	10	20	19	60	77			
FUR1, 163-172 (170)								A	22	21	29	32	47	73	62			
ADE3, 603-621 (612)								E	14	18	13	13	40	50	59			
TEF1, 356-369 (361;368)								F	18	16	6	7	16	18	14			

REFERENCES

1. Dreveny I, Pye VE, Beuron F, Briggs LC, Isaacson RL, et al. (2004) p97 and close encounters of every kind: a brief review. *Biochemical Society transactions* **32**(Pt 5): 715-720.
2. Hampton MB, Kettle AJ, Winterbourn CC (1998) Inside the neutrophil phagosome: oxidants, myeloperoxidase, and bacterial killing. *Blood* **92**(9): 3007-3017.
3. Cadenas E, Davies KJ (2000) Mitochondrial free radical generation, oxidative stress, and aging. *Free radical biology & medicine* **29**(3-4): 222-230.
4. Suh YA, Arnold RS, Lassegue B, Shi J, Xu X, et al. (1999) Cell transformation by the superoxide-generating oxidase Mox1. *Nature* **401**(6748): 79-82.
5. Finkel T, Holbrook NJ (2000) Oxidants, oxidative stress and the biology of ageing. *Nature* **408**(6809): 239-247.
6. D'Autreaux B, Toledano MB (2007) ROS as signalling molecules: mechanisms that generate specificity in ROS homeostasis. *Nature reviews* **8**(10): 813-824.
7. Seth D, Rudolph J (2006) Redox regulation of MAP kinase phosphatase 3. *Biochemistry* **45**(28): 8476-8487.
8. Stadtman ER, Berlett BS (1998) Reactive oxygen-mediated protein oxidation in aging and disease. *Drug metabolism reviews* **30**(2): 225-243.
9. Moncada S, Palmer RM, Higgs EA (1991) Nitric oxide: physiology, pathophysiology, and pharmacology. *Pharmacological reviews* **43**(2): 109-142.
10. Schrammel A, Gorren AC, Schmidt K, Pfeiffer S, Mayer B (2003) S-nitrosation of glutathione by nitric oxide, peroxynitrite, and (*)NO/O(2)(*-). *Free radical biology & medicine* **34**(8): 1078-1088.
11. Hess DT, Matsumoto A, Kim SO, Marshall HE, Stamler JS (2005) Protein S-nitrosylation: purview and parameters. *Nature reviews* **6**(2): 150-166.
12. Rhee SG, Chae HZ, Kim K (2005) Peroxiredoxins: a historical overview and speculative preview of novel mechanisms and emerging concepts in cell signaling. *Free radical biology & medicine* **38**(12): 1543-1552.
13. Holmgren A, Johansson C, Berndt C, Lonn ME, Hudemann C, et al. (2005) Thiol redox control via thioredoxin and glutaredoxin systems. *Biochemical Society transactions* **33**(Pt 6): 1375-1377.
14. Jones DP (2002) Redox potential of GSH/GSSG couple: assay and biological significance. *Methods in enzymology* **348**: 93-112.
15. Kirlin WG, Cai J, Thompson SA, Diaz D, Kavanagh TJ, et al. (1999) Glutathione redox potential in response to differentiation and enzyme inducers. *Free radical biology & medicine* **27**(11-12): 1208-1218.
16. Tajc SG, Tolbert BS, Basavappa R, Miller BL (2004) Direct determination of thiol pKa by isothermal titration microcalorimetry. *Journal of the American Chemical Society* **126**(34): 10508-10509.
17. Berndt C, Lillig CH, Holmgren A (2007) Thiol-based mechanisms of the thioredoxin and glutaredoxin systems: implications for diseases in the cardiovascular system. *Am J Physiol Heart Circ Physiol* **292**(3): H1227-1236.
18. Marnett LJ (2000) Oxyradicals and DNA damage. *Carcinogenesis* **21**(3): 361-370.
19. Bartsch H, Nair J (2004) Oxidative stress and lipid peroxidation-derived DNA-lesions in inflammation driven carcinogenesis. *Cancer detection and prevention* **28**(6): 385-391.
20. Sies H (1991) Role of reactive oxygen species in biological processes. *Klinische Wochenschrift* **69**(21-23): 965-968.
21. Gasch AP, Spellman PT, Kao CM, Carmel-Harel O, Eisen MB, et al. (2000) Genomic expression programs in the response of yeast cells to environmental changes. *Molecular biology of the cell* **11**(12): 4241-4257.

22. Lowell BB, Shulman GI (2005) Mitochondrial dysfunction and type 2 diabetes. *Science (New York, NY* **307**(5708): 384-387.
23. Cook JA, Gius D, Wink DA, Krishna MC, Russo A, et al. (2004) Oxidative stress, redox, and the tumor microenvironment. *Seminars in radiation oncology* **14**(3): 259-266.
24. Barford D (2004) The role of cysteine residues as redox-sensitive regulatory switches. *Current opinion in structural biology* **14**(6): 679-686.
25. Lee SR, Kwon KS, Kim SR, Rhee SG (1998) Reversible inactivation of protein-tyrosine phosphatase 1B in A431 cells stimulated with epidermal growth factor. *The Journal of biological chemistry* **273**(25): 15366-15372.
26. Yu CX, Li S, Whorton AR (2005) Redox regulation of PTEN by S-nitrosothiols. *Molecular pharmacology* **68**(3): 847-854.
27. Rainwater R, Parks D, Anderson ME, Tegtmeyer P, Mann K (1995) Role of cysteine residues in regulation of p53 function. *Molecular and cellular biology* **15**(7): 3892-3903.
28. Ahn SG, Thiele DJ (2003) Redox regulation of mammalian heat shock factor 1 is essential for Hsp gene activation and protection from stress. *Genes & development* **17**(4): 516-528.
29. Ilbert M, Horst J, Ahrens S, Winter J, Graf PC, et al. (2007) The redox-switch domain of Hsp33 functions as dual stress sensor. *Nature structural & molecular biology* **14**(6): 556-563.
30. Giles GI, Tasker KM, Jacob C (2001) Hypothesis: the role of reactive sulfur species in oxidative stress. *Free radical biology & medicine* **31**(10): 1279-1283.
31. Lohse DL, Denu JM, Santoro N, Dixon JE (1997) Roles of aspartic acid-181 and serine-222 in intermediate formation and hydrolysis of the mammalian protein-tyrosine-phosphatase PTP1. *Biochemistry* **36**(15): 4568-4575.
32. Ma LH, Takanishi CL, Wood MJ (2007) Molecular mechanism of oxidative stress perception by the Orp1 protein. *The Journal of biological chemistry* **282**(43): 31429-31436.
33. Rhee SG, Bae YS, Lee SR, Kwon J (2000) Hydrogen peroxide: a key messenger that modulates protein phosphorylation through cysteine oxidation. *Sci STKE* **2000**(53): PE1.
34. Zhang ZY, Dixon JE (1993) Active site labeling of the Yersinia protein tyrosine phosphatase: the determination of the pKa of the active site cysteine and the function of the conserved histidine 402. *Biochemistry* **32**(36): 9340-9345.
35. Winterbourn CC, Metodiewa D (1999) Reactivity of biologically important thiol compounds with superoxide and hydrogen peroxide. *Free radical biology & medicine* **27**(3-4): 322-328.
36. Stamler JS (1994) Redox signaling: nitrosylation and related target interactions of nitric oxide. *Cell* **78**(6): 931-936.
37. Cumming RC, Andon NL, Haynes PA, Park M, Fischer WH, et al. (2004) Protein disulfide bond formation in the cytoplasm during oxidative stress. *The Journal of biological chemistry* **279**(21): 21749-21758.
38. Salmeen A, Andersen JN, Myers MP, Meng TC, Hinks JA, et al. (2003) Redox regulation of protein tyrosine phosphatase 1B involves a sulphenyl-amide intermediate. *Nature* **423**(6941): 769-773.
39. Dalle-Donne I, Rossi R, Giustarini D, Colombo R, Milzani A (2007) S-glutathionylation in protein redox regulation. *Free radical biology & medicine* **43**(6): 883-898.
40. Cotgreave IA, Gerdes R, Schuppe-Koistinen I, Lind C (2002) S-glutathionylation of glyceraldehyde-3-phosphate dehydrogenase: role of thiol oxidation and catalysis by glutaredoxin. *Methods in enzymology* **348**: 175-182.
41. Shelton MD, Chock PB, Mieyal JJ (2005) Glutaredoxin: role in reversible protein s-glutathionylation and regulation of redox signal transduction and protein translocation. *Antioxidants & redox signaling* **7**(3-4): 348-366.
42. Leichert LI, Jakob U (2004) Protein thiol modifications visualized in vivo. *PLoS biology* **2**(11): e333.

43. Aslund F, Zheng M, Beckwith J, Storz G (1999) Regulation of the OxyR transcription factor by hydrogen peroxide and the cellular thiol-disulfide status. *Proceedings of the National Academy of Sciences of the United States of America* **96**(11): 6161-6165.
44. Mallis RJ, Hamann MJ, Zhao W, Zhang T, Hendrich S, et al. (2002) Irreversible thiol oxidation in carbonic anhydrase III: protection by S-glutathiolation and detection in aging rats. *Biological chemistry* **383**(3-4): 649-662.
45. Biteau B, Labarre J, Toledano MB (2003) ATP-dependent reduction of cysteine-sulphinic acid by S. cerevisiae sulphiredoxin. *Nature* **425**(6961): 980-984.
46. Woo HA, Chae HZ, Hwang SC, Yang KS, Kang SW, et al. (2003) Reversing the inactivation of peroxiredoxins caused by cysteine sulfinic acid formation. *Science (New York, NY* **300**(5619): 653-656.
47. Sitia R, Molteni SN (2004) Stress, protein (mis)folding, and signaling: the redox connection. *Sci STKE* **2004**(239): pe27.
48. Cumming RC, Schubert D (2005) Amyloid-beta induces disulfide bonding and aggregation of GAPDH in Alzheimer's disease. *Faseb J* **19**(14): 2060-2062.
49. Leichert LI, Gehrke F, Gudiseva HV, Blackwell T, Ilbert M, et al. (2008) Quantifying changes in the thiol redox proteome upon oxidative stress in vivo. *Proceedings of the National Academy of Sciences of the United States of America* **105**(24): 8197-8202.
50. Paget MS, Buttner MJ (2003) Thiol-based regulatory switches. *Annual review of genetics* **37**: 91-121.
51. Shenton D, Grant CM (2003) Protein S-thiolation targets glycolysis and protein synthesis in response to oxidative stress in the yeast Saccharomyces cerevisiae. *The Biochemical journal* **374**(Pt 2): 513-519.
52. Landino LM (2008) Protein thiol modification by peroxynitrite anion and nitric oxide donors. *Methods in enzymology* **440**: 95-109.
53. Ying J, Clavreul N, Sethuraman M, Adachi T, Cohen RA (2007) Thiol oxidation in signaling and response to stress: detection and quantification of physiological and pathophysiological thiol modifications. *Free radical biology & medicine* **43**(8): 1099-1108.
54. Leichert LI, Jakob U (2006) Global methods to monitor the thiol-disulfide state of proteins in vivo. *Antioxidants & redox signaling* **8**(5-6): 763-772.
55. Kamata H, Hirata H (1999) Redox regulation of cellular signalling. *Cellular signalling* **11**(1): 1-14.
56. Allen RG, Tresini M (2000) Oxidative stress and gene regulation. *Free radical biology & medicine* **28**(3): 463-499.
57. Meyer M, Schreck R, Baeuerle PA (1993) H2O2 and antioxidants have opposite effects on activation of NF-kappa B and AP-1 in intact cells: AP-1 as secondary antioxidant-responsive factor. *The EMBO journal* **12**(5): 2005-2015.
58. Bandyopadhyay S, Gronostajski RM (1994) Identification of a conserved oxidation-sensitive cysteine residue in the NFI family of DNA-binding proteins. *The Journal of biological chemistry* **269**(47): 29949-29955.
59. Pognonec P, Kato H, Roeder RG (1992) The helix-loop-helix/leucine repeat transcription factor USF can be functionally regulated in a redox-dependent manner. *The Journal of biological chemistry* **267**(34): 24563-24567.
60. Karin M, Liu Z, Zandi E (1997) AP-1 function and regulation. *Current opinion in cell biology* **9**(2): 240-246.
61. Izawa S, Maeda K, Sugiyama K, Mano J, Inoue Y, et al. (1999) Thioredoxin deficiency causes the constitutive activation of Yap1, an AP-1-like transcription factor in Saccharomyces cerevisiae. *The Journal of biological chemistry* **274**(40): 28459-28465.
62. Toone WM, Morgan BA, Jones N (2001) Redox control of AP-1-like factors in yeast and beyond. *Oncogene* **20**(19): 2336-2346.

63. Kuge S, Jones N (1994) YAP1 dependent activation of TRX2 is essential for the response of Saccharomyces cerevisiae to oxidative stress by hydroperoxides. *The EMBO journal* **13**(3): 655-664.
64. Nguyen DT, Alarco AM, Raymond M (2001) Multiple Yap1p-binding sites mediate induction of the yeast major facilitator FLR1 gene in response to drugs, oxidants, and alkylating agents. *The Journal of biological chemistry* **276**(2): 1138-1145.
65. Azevedo D, Tacnet F, Delaunay A, Rodrigues-Pousada C, Toledano MB (2003) Two redox centers within Yap1 for H2O2 and thiol-reactive chemicals signaling. *Free radical biology & medicine* **35**(8): 889-900.
66. Inoue Y, Matsuda T, Sugiyama K, Izawa S, Kimura A (1999) Genetic analysis of glutathione peroxidase in oxidative stress response of Saccharomyces cerevisiae. *The Journal of biological chemistry* **274**(38): 27002-27009.
67. Carmel-Harel O, Stearman R, Gasch AP, Botstein D, Brown PO, et al. (2001) Role of thioredoxin reductase in the Yap1p-dependent response to oxidative stress in Saccharomyces cerevisiae. *Molecular microbiology* **39**(3): 595-605.
68. Delaunay A, Isnard AD, Toledano MB (2000) H2O2 sensing through oxidation of the Yap1 transcription factor. *The EMBO journal* **19**(19): 5157-5166.
69. Wood MJ, Storz G, Tjandra N (2004) Structural basis for redox regulation of Yap1 transcription factor localization. *Nature* **430**(7002): 917-921.
70. Kuge S, Jones N, Nomoto A (1997) Regulation of yAP-1 nuclear localization in response to oxidative stress. *The EMBO journal* **16**(7): 1710-1720.
71. Yan C, Lee LH, Davis LI (1998) Crm1p mediates regulated nuclear export of a yeast AP-1-like transcription factor. *The EMBO journal* **17**(24): 7416-7429.
72. Kuge S, Toda T, Iizuka N, Nomoto A (1998) Crm1 (XpoI) dependent nuclear export of the budding yeast transcription factor yAP-1 is sensitive to oxidative stress. *Genes Cells* **3**(8): 521-532.
73. Delaunay A, Pflieger D, Barrault MB, Vinh J, Toledano MB (2002) A thiol peroxidase is an H2O2 receptor and redox-transducer in gene activation. *Cell* **111**(4): 471-481.
74. Okazaki S, Tachibana T, Naganuma A, Mano N, Kuge S (2007) Multistep disulfide bond formation in Yap1 is required for sensing and transduction of H2O2 stress signal. *Molecular cell* **27**(4): 675-688.
75. Xanthoudakis S, Curran T (1992) Identification and characterization of Ref-1, a nuclear protein that facilitates AP-1 DNA-binding activity. *The EMBO journal* **11**(2): 653-665.
76. Walker LJ, Robson CN, Black E, Gillespie D, Hickson ID (1993) Identification of residues in the human DNA repair enzyme HAP1 (Ref-1) that are essential for redox regulation of Jun DNA binding. *Molecular and cellular biology* **13**(9): 5370-5376.
77. Abate C, Patel L, Rauscher FJ, 3rd, Curran T (1990) Redox regulation of fos and jun DNA-binding activity in vitro. *Science (New York, NY* **249**(4973): 1157-1161.
78. Okuno H, Akahori A, Sato H, Xanthoudakis S, Curran T, et al. (1993) Escape from redox regulation enhances the transforming activity of Fos. *Oncogene* **8**(3): 695-701.
79. Xanthoudakis S, Miao GG, Curran T (1994) The redox and DNA-repair activities of Ref-1 are encoded by nonoverlapping domains. *Proceedings of the National Academy of Sciences of the United States of America* **91**(1): 23-27.
80. Gulshan K, Rovinsky SA, Coleman ST, Moye-Rowley WS (2005) Oxidant-specific folding of Yap1p regulates both transcriptional activation and nuclear localization. *The Journal of biological chemistry* **280**(49): 40524-40533.
81. Veal EA, Ross SJ, Malakasi P, Peacock E, Morgan BA (2003) Ybp1 is required for the hydrogen peroxide-induced oxidation of the Yap1 transcription factor. *The Journal of biological chemistry* **278**(33): 30896-30904.

82. Okazaki S, Naganuma A, Kuge S (2005) Peroxiredoxin-mediated redox regulation of the nuclear localization of Yap1, a transcription factor in budding yeast. *Antioxidants & redox signaling* **7**(3-4): 327-334.
83. Kuge S, Arita M, Murayama A, Maeta K, Izawa S, et al. (2001) Regulation of the yeast Yap1p nuclear export signal is mediated by redox signal-induced reversible disulfide bond formation. *Molecular and cellular biology* **21**(18): 6139-6150.
84. Castillo EA, Ayte J, Chiva C, Moldon A, Carrascal M, et al. (2002) Diethylmaleate activates the transcription factor Pap1 by covalent modification of critical cysteine residues. *Molecular microbiology* **45**(1): 243-254.
85. Paget MS, Bae JB, Hahn MY, Li W, Kleanthous C, et al. (2001) Mutational analysis of RsrA, a zinc-binding anti-sigma factor with a thiol-disulphide redox switch. *Molecular microbiology* **39**(4): 1036-1047.
86. Vivancos AP, Castillo EA, Biteau B, Nicot C, Ayte J, et al. (2005) A cysteine-sulfinic acid in peroxiredoxin regulates H2O2-sensing by the antioxidant Pap1 pathway. *Proceedings of the National Academy of Sciences of the United States of America* **102**(25): 8875-8880.
87. Dinkova-Kostova AT, Holtzclaw WD, Kensler TW (2005) The role of Keap1 in cellular protective responses. *Chemical research in toxicology* **18**(12): 1779-1791.
88. Kobayashi M, Yamamoto M (2006) Nrf2-Keap1 regulation of cellular defense mechanisms against electrophiles and reactive oxygen species. *Advances in enzyme regulation* **46**: 113-140.
89. Kobayashi A, Kang MI, Okawa H, Ohtsuji M, Zenke Y, et al. (2004) Oxidative stress sensor Keap1 functions as an adaptor for Cul3-based E3 ligase to regulate proteasomal degradation of Nrf2. *Molecular and cellular biology* **24**(16): 7130-7139.
90. Zhang DD (2006) Mechanistic studies of the Nrf2-Keap1 signaling pathway. *Drug metabolism reviews* **38**(4): 769-789.
91. Nguyen T, Sherratt PJ, Pickett CB (2003) Regulatory mechanisms controlling gene expression mediated by the antioxidant response element. *Annual review of pharmacology and toxicology* **43**: 233-260.
92. Reddy S, Jones AD, Cross CE, Wong PS, Van Der Vliet A (2000) Inactivation of creatine kinase by S-glutathionylation of the active-site cysteine residue. *The Biochemical journal* **347 Pt 3**: 821-827.
93. Kim G, Levine RL (2005) Molecular determinants of S-glutathionylation of carbonic anhydrase 3. *Antioxidants & redox signaling* **7**(7-8): 849-854.
94. Mohr S, Hallak H, de Boitte A, Lapetina EG, Brune B (1999) Nitric oxide-induced S-glutathionylation and inactivation of glyceraldehyde-3-phosphate dehydrogenase. *The Journal of biological chemistry* **274**(14): 9427-9430.
95. Brodie AE, Reed DJ (1990) Cellular recovery of glyceraldehyde-3-phosphate dehydrogenase activity and thiol status after exposure to hydroperoxides. *Archives of biochemistry and biophysics* **276**(1): 212-218.
96. Sirover MA (1999) New insights into an old protein: the functional diversity of mammalian glyceraldehyde-3-phosphate dehydrogenase. *Biochimica et biophysica acta* **1432**(2): 159-184.
97. Zheng L, Roeder RG, Luo Y (2003) S phase activation of the histone H2B promoter by OCA-S, a coactivator complex that contains GAPDH as a key component. *Cell* **114**(2): 255-266.
98. Patterson RL, van Rossum DB, Kaplin AI, Barrow RK, Snyder SH (2005) Inositol 1,4,5-trisphosphate receptor/GAPDH complex augments Ca2+ release via locally derived NADH. *Proceedings of the National Academy of Sciences of the United States of America* **102**(5): 1357-1359.
99. Chuang DM, Hough C, Senatorov VV (2005) Glyceraldehyde-3-phosphate dehydrogenase, apoptosis, and neurodegenerative diseases. *Annual review of pharmacology and toxicology* **45**: 269-290.

100. Hara MR, Agrawal N, Kim SF, Cascio MB, Fujimuro M, et al. (2005) S-nitrosylated GAPDH initiates apoptotic cell death by nuclear translocation following Siah1 binding. *Nature cell biology* **7**(7): 665-674.
101. Senatorov VV, Charles V, Reddy PH, Tagle DA, Chuang DM (2003) Overexpression and nuclear accumulation of glyceraldehyde-3-phosphate dehydrogenase in a transgenic mouse model of Huntington's disease. *Molecular and cellular neurosciences* **22**(3): 285-297.
102. Tsuchiya K, Tajima H, Kuwae T, Takeshima T, Nakano T, et al. (2005) Pro-apoptotic protein glyceraldehyde-3-phosphate dehydrogenase promotes the formation of Lewy body-like inclusions. *The European journal of neuroscience* **21**(2): 317-326.
103. Mazzola JL, Sirover MA (2003) Subcellular alteration of glyceraldehyde-3-phosphate dehydrogenase in Alzheimer's disease fibroblasts. *Journal of neuroscience research* **71**(2): 279-285.
104. Sawa A, Khan AA, Hester LD, Snyder SH (1997) Glyceraldehyde-3-phosphate dehydrogenase: nuclear translocation participates in neuronal and nonneuronal cell death. *Proceedings of the National Academy of Sciences of the United States of America* **94**(21): 11669-11674.
105. Cahuana GM, Tejedo JR, Jimenez J, Ramirez R, Sobrino F, et al. (2004) Nitric oxide-induced carbonylation of Bcl-2, GAPDH and ANT precedes apoptotic events in insulin-secreting RINm5F cells. *Experimental cell research* **293**(1): 22-30.
106. Dimmeler S, Lottspeich F, Brune B (1992) Nitric oxide causes ADP-ribosylation and inhibition of glyceraldehyde-3-phosphate dehydrogenase. *The Journal of biological chemistry* **267**(24): 16771-16774.
107. Ralser M, Wamelink MM, Kowald A, Gerisch B, Heeren G, et al. (2007) Dynamic rerouting of the carbohydrate flux is key to counteracting oxidative stress. *J Biol* **6**(4): 10.
108. Godon C, Lagniel G, Lee J, Buhler JM, Kieffer S, et al. (1998) The H2O2 stimulon in Saccharomyces cerevisiae. *The Journal of biological chemistry* **273**(35): 22480-22489.
109. Arner ES, Holmgren A (2000) Physiological functions of thioredoxin and thioredoxin reductase. *European journal of biochemistry / FEBS* **267**(20): 6102-6109.
110. Ralser M, Heeren G, Breitenbach M, Lehrach H, Krobitsch S (2006) Triose phosphate isomerase deficiency is caused by altered dimerization--not catalytic inactivity--of the mutant enzymes. *PLoS ONE* **1**: e30.
111. Kuo SC, Lampen JO (1972) Inhibition by 2-deoxy-D-glucose of synthesis of glycoprotein enzymes by protoplasts of Saccharomyces: relation to inhibition of sugar uptake and metabolism. *Journal of bacteriology* **111**(2): 419-429.
112. Nakajima H, Amano W, Fujita A, Fukuhara A, Azuma YT, et al. (2007) The active site cysteine of the proapoptotic protein glyceraldehyde-3-phosphate dehydrogenase is essential in oxidative stress-induced aggregation and cell death. *The Journal of biological chemistry* **282**(36): 26562-26574.
113. Krishnan S, Chi EY, Wood SJ, Kendrick BS, Li C, et al. (2003) Oxidative dimer formation is the critical rate-limiting step for Parkinson's disease alpha-synuclein fibrillogenesis. *Biochemistry* **42**(3): 829-837.
114. Benzinger TL, Gregory DM, Burkoth TS, Miller-Auer H, Lynn DG, et al. (2000) Two-dimensional structure of beta-amyloid(10-35) fibrils. *Biochemistry* **39**(12): 3491-3499.
115. Ross CA, Poirier MA (2004) Protein aggregation and neurodegenerative disease. *Nature medicine* **10 Suppl**: S10-17.
116. Nagai Y, Inui T, Popiel HA, Fujikake N, Hasegawa K, et al. (2007) A toxic monomeric conformer of the polyglutamine protein. *Nature structural & molecular biology* **14**(4): 332-340.
117. House CM, Frew IJ, Huang HL, Wiche G, Traficante N, et al. (2003) A binding motif for Siah ubiquitin ligase. *Proceedings of the National Academy of Sciences of the United States of America* **100**(6): 3101-3106.

118. Roperch JP, Lethrone F, Prieur S, Piouffre L, Israeli D, et al. (1999) SIAH-1 promotes apoptosis and tumor suppression through a network involving the regulation of protein folding, unfolding, and trafficking: identification of common effectors with p53 and p21(Waf1). *Proceedings of the National Academy of Sciences of the United States of America* **96**(14): 8070-8073.
119. Ronai Z (1993) Glycolytic enzymes as DNA binding proteins. *The International journal of biochemistry* **25**(7): 1073-1076.
120. Shi Y, Shi Y (2004) Metabolic enzymes and coenzymes in transcription--a direct link between metabolism and transcription? *Trends Genet* **20**(9): 445-452.
121. Shi Y, Sawada J, Sui G, Affar el B, Whetstine JR, et al. (2003) Coordinated histone modifications mediated by a CtBP co-repressor complex. *Nature* **422**(6933): 735-738.
122. Morigasaki S, Shimada K, Ikner A, Yanagida M, Shiozaki K (2008) Glycolytic enzyme GAPDH promotes peroxide stress signaling through multistep phosphorelay to a MAPK cascade. *Molecular cell* **30**(1): 108-113.
123. Saito H (2001) Histidine phosphorylation and two-component signaling in eukaryotic cells. *Chemical reviews* **101**(8): 2497-2509.
124. Shiozaki K, Shiozaki M, Russell P (1998) Heat stress activates fission yeast Spc1/Sty1 MAPK by a MEKK-independent mechanism. *Molecular biology of the cell* **9**(6): 1339-1349.
125. Wilkinson MG, Samuels M, Takeda T, Toone WM, Shieh JC, et al. (1996) The Atf1 transcription factor is a target for the Sty1 stress-activated MAP kinase pathway in fission yeast. *Genes & development* **10**(18): 2289-2301.
126. Aoyama K, Mitsubayashi Y, Aiba H, Mizuno T (2000) Spy1, a histidine-containing phosphotransfer signaling protein, regulates the fission yeast cell cycle through the Mcs4 response regulator. *Journal of bacteriology* **182**(17): 4868-4874.
127. Ziegler DM (1985) Role of reversible oxidation-reduction of enzyme thiols-disulfides in metabolic regulation. *Annual review of biochemistry* **54**: 305-329.
128. Steeghs K, Benders A, Oerlemans F, de Haan A, Heerschap A, et al. (1997) Altered Ca2+ responses in muscles with combined mitochondrial and cytosolic creatine kinase deficiencies. *Cell* **89**(1): 93-103.
129. Schlattner U, Tokarska-Schlattner M, Wallimann T (2006) Mitochondrial creatine kinase in human health and disease. *Biochimica et biophysica acta* **1762**(2): 164-180.
130. Ernest MJ, Kim KH (1974) Regulation of rat liver glycogen synthetase D. Role of glucose 6-phosphate and enzyme sulfhydryl groups in activity and glycogen binding. *The Journal of biological chemistry* **249**(16): 5011-5018.
131. Shimazu T, Tokutake S, Usami M (1978) Inactivation of phosphorylase phosphatase by a factor from rabbit liver and its chemical characterization as glutathione disulfide. *The Journal of biological chemistry* **253**(20): 7376-7382.
132. Cabiscol E, Levine RL (1995) Carbonic anhydrase III. Oxidative modification in vivo and loss of phosphatase activity during aging. *The Journal of biological chemistry* **270**(24): 14742-14747.
133. Alonso A, Sasin J, Bottini N, Friedberg I, Friedberg I, et al. (2004) Protein tyrosine phosphatases in the human genome. *Cell* **117**(6): 699-711.
134. Barford D, Das AK, Egloff MP (1998) The structure and mechanism of protein phosphatases: insights into catalysis and regulation. *Annual review of biophysics and biomolecular structure* **27**: 133-164.
135. Barford D, Flint AJ, Tonks NK (1994) Crystal structure of human protein tyrosine phosphatase 1B. *Science (New York, NY* **263**(5152): 1397-1404.
136. Fauman EB, Saper MA (1996) Structure and function of the protein tyrosine phosphatases. *Trends in biochemical sciences* **21**(11): 413-417.
137. Finkel T (1998) Oxygen radicals and signaling. *Current opinion in cell biology* **10**(2): 248-253.

138. Denu JM, Tanner KG (1998) Specific and reversible inactivation of protein tyrosine phosphatases by hydrogen peroxide: evidence for a sulfenic acid intermediate and implications for redox regulation. *Biochemistry* **37**(16): 5633-5642.
139. Dube N, Tremblay ML (2005) Involvement of the small protein tyrosine phosphatases TC-PTP and PTP1B in signal transduction and diseases: from diabetes, obesity to cell cycle, and cancer. *Biochimica et biophysica acta* **1754**(1-2): 108-117.
140. Elchebly M, Payette P, Michaliszyn E, Cromlish W, Collins S, et al. (1999) Increased insulin sensitivity and obesity resistance in mice lacking the protein tyrosine phosphatase-1B gene. *Science (New York, NY* **283**(5407): 1544-1548.
141. Zhu S, Bjorge JD, Fujita DJ (2007) PTP1B contributes to the oncogenic properties of colon cancer cells through Src activation. *Cancer research* **67**(21): 10129-10137.
142. van Montfort RL, Congreve M, Tisi D, Carr R, Jhoti H (2003) Oxidation state of the active-site cysteine in protein tyrosine phosphatase 1B. *Nature* **423**(6941): 773-777.
143. Lee JW, Soonsanga S, Helmann JD (2007) A complex thiolate switch regulates the Bacillus subtilis organic peroxide sensor OhrR. *Proceedings of the National Academy of Sciences of the United States of America* **104**(21): 8743-8748.
144. Barrett WC, DeGnore JP, Konig S, Fales HM, Keng YF, et al. (1999) Regulation of PTP1B via glutathionylation of the active site cysteine 215. *Biochemistry* **38**(20): 6699-6705.
145. Salmeen A, Barford D (2005) Functions and mechanisms of redox regulation of cysteine-based phosphatases. *Antioxidants & redox signaling* **7**(5-6): 560-577.
146. Lou YW, Chen YY, Hsu SF, Chen RK, Lee CL, et al. (2008) Redox regulation of the protein tyrosine phosphatase PTP1B in cancer cells. *The FEBS journal* **275**(1): 69-88.
147. Huang ZZ, Chen C, Zeng Z, Yang H, Oh J, et al. (2001) Mechanism and significance of increased glutathione level in human hepatocellular carcinoma and liver regeneration. *Faseb J* **15**(1): 19-21.
148. Wood ZA, Poole LB, Hantgan RR, Karplus PA (2002) Dimers to doughnuts: redox-sensitive oligomerization of 2-cysteine peroxiredoxins. *Biochemistry* **41**(17): 5493-5504.
149. Jang HH, Lee KO, Chi YH, Jung BG, Park SK, et al. (2004) Two enzymes in one; two yeast peroxiredoxins display oxidative stress-dependent switching from a peroxidase to a molecular chaperone function. *Cell* **117**(5): 625-635.
150. Yang J, Groen A, Lemeer S, Jans A, Slijper M, et al. (2007) Reversible oxidation of the membrane distal domain of receptor PTPalpha is mediated by a cyclic sulfenamide. *Biochemistry* **46**(3): 709-719.
151. Meng TC, Fukada T, Tonks NK (2002) Reversible oxidation and inactivation of protein tyrosine phosphatases in vivo. *Molecular cell* **9**(2): 387-399.
152. Kamata H, Honda S, Maeda S, Chang L, Hirata H, et al. (2005) Reactive oxygen species promote TNFalpha-induced death and sustained JNK activation by inhibiting MAP kinase phosphatases. *Cell* **120**(5): 649-661.
153. Sun H, Lesche R, Li DM, Liliental J, Zhang H, et al. (1999) PTEN modulates cell cycle progression and cell survival by regulating phosphatidylinositol 3,4,5,-trisphosphate and Akt/protein kinase B signaling pathway. *Proceedings of the National Academy of Sciences of the United States of America* **96**(11): 6199-6204.
154. Messens J, Hayburn G, Desmyter A, Laus G, Wyns L (1999) The essential catalytic redox couple in arsenate reductase from Staphylococcus aureus. *Biochemistry* **38**(51): 16857-16865.
155. Meng TC, Buckley DA, Galic S, Tiganis T, Tonks NK (2004) Regulation of insulin signaling through reversible oxidation of the protein-tyrosine phosphatases TC45 and PTP1B. *The Journal of biological chemistry* **279**(36): 37716-37725.
156. Imam A, Hoyos B, Swenson C, Levi E, Chua R, et al. (2001) Retinoids as ligands and coactivators of protein kinase C alpha. *Faseb J* **15**(1): 28-30.
157. Jakob U, Muse W, Eser M, Bardwell JC (1999) Chaperone activity with a redox switch. *Cell* **96**(3): 341-352.

158. Brandes N, Rinck A, Leichert LI, Jakob U (2007) Nitrosative stress treatment of E. coli targets distinct set of thiol-containing proteins. *Molecular microbiology* **66**(4): 901-914.
159. Ji XB, Hollocher TC (1988) Reduction of nitrite to nitric oxide by enteric bacteria. *Biochemical and biophysical research communications* **157**(1): 106-108.
160. Fang FC (1997) Perspectives series: host/pathogen interactions. Mechanisms of nitric oxide-related antimicrobial activity. *The Journal of clinical investigation* **99**(12): 2818-2825.
161. Nathan C (1992) Nitric oxide as a secretory product of mammalian cells. *Faseb J* **6**(12): 3051-3064.
162. Poole RK (2005) Nitric oxide and nitrosative stress tolerance in bacteria. *Biochemical Society transactions* **33**(Pt 1): 176-180.
163. Stamler JS, Singel DJ, Loscalzo J (1992) Biochemistry of nitric oxide and its redox-activated forms. *Science (New York, NY* **258**(5090): 1898-1902.
164. Kim YM, Son K, Hong SJ, Green A, Chen JJ, et al. (1998) Inhibition of protein synthesis by nitric oxide correlates with cytostatic activity: nitric oxide induces phosphorylation of initiation factor eIF-2 alpha. *Molecular medicine (Cambridge, Mass* **4**(3): 179-190.
165. Bundy RE, Marczin N, Chester AH, Yacoub M (2000) A redox-based mechanism for nitric oxide-induced inhibition of DNA synthesis in human vascular smooth muscle cells. *British journal of pharmacology* **129**(7): 1513-1521.
166. Beckman JS (1996) Oxidative damage and tyrosine nitration from peroxynitrite. *Chemical research in toxicology* **9**(5): 836-844.
167. Arnold WP, Mittal CK, Katsuki S, Murad F (1977) Nitric oxide activates guanylate cyclase and increases guanosine 3':5'-cyclic monophosphate levels in various tissue preparations. *Proceedings of the National Academy of Sciences of the United States of America* **74**(8): 3203-3207.
168. Nunoshiba T, deRojas-Walker T, Wishnok JS, Tannenbaum SR, Demple B (1993) Activation by nitric oxide of an oxidative-stress response that defends Escherichia coli against activated macrophages. *Proceedings of the National Academy of Sciences of the United States of America* **90**(21): 9993-9997.
169. Cruz-Ramos H, Crack J, Wu G, Hughes MN, Scott C, et al. (2002) NO sensing by FNR: regulation of the Escherichia coli NO-detoxifying flavohaemoglobin, Hmp. *The EMBO journal* **21**(13): 3235-3244.
170. Sun J, Steenbergen C, Murphy E (2006) S-nitrosylation: NO-related redox signaling to protect against oxidative stress. *Antioxidants & redox signaling* **8**(9-10): 1693-1705.
171. Poole LB, Karplus PA, Claiborne A (2004) Protein sulfenic acids in redox signaling. *Annual review of pharmacology and toxicology* **44**: 325-347.
172. Ishii T, Sunami O, Nakajima H, Nishio H, Takeuchi T, et al. (1999) Critical role of sulfenic acid formation of thiols in the inactivation of glyceraldehyde-3-phosphate dehydrogenase by nitric oxide. *Biochemical pharmacology* **58**(1): 133-143.
173. Rhee KY, Erdjument-Bromage H, Tempst P, Nathan CF (2005) S-nitroso proteome of Mycobacterium tuberculosis: Enzymes of intermediary metabolism and antioxidant defense. *Proceedings of the National Academy of Sciences of the United States of America* **102**(2): 467-472.
174. Kang Y, Durfee T, Glasner JD, Qiu Y, Frisch D, et al. (2004) Systematic mutagenesis of the Escherichia coli genome. *Journal of bacteriology* **186**(15): 4921-4930.
175. Neidhardt FC, Bloch PL, Smith DF (1974) Culture medium for enterobacteria. *Journal of bacteriology* **119**(3): 736-747.
176. Maragos CM, Morley D, Wink DA, Dunams TM, Saavedra JE, et al. (1991) Complexes of .NO with nucleophiles as agents for the controlled biological release of nitric oxide. Vasorelaxant effects. *Journal of medicinal chemistry* **34**(11): 3242-3247.

177. Dykhuizen RS, Frazer R, Duncan C, Smith CC, Golden M, et al. (1996) Antimicrobial effect of acidified nitrite on gut pathogens: importance of dietary nitrate in host defense. *Antimicrobial agents and chemotherapy* **40**(6): 1422-1425.
178. Hiniker A, Bardwell JC (2004) In vivo substrate specificity of periplasmic disulfide oxidoreductases. *The Journal of biological chemistry* **279**(13): 12967-12973.
179. Hashim S, Kwon DH, Abdelal A, Lu CD (2004) The arginine regulatory protein mediates repression by arginine of the operons encoding glutamate synthase and anabolic glutamate dehydrogenase in Pseudomonas aeruginosa. *Journal of bacteriology* **186**(12): 3848-3854.
180. Goss TJ, Perez-Matos A, Bender RA (2001) Roles of glutamate synthase, gltBD, and gltF in nitrogen metabolism of Escherichia coli and Klebsiella aerogenes. *Journal of bacteriology* **183**(22): 6607-6619.
181. Meers JL, Tempest DW, Brown CM (1970) 'Glutamine(amide):2-oxoglutarate amino transferase oxido-reductase (NADP); an enzyme involved in the synthesis of glutamate by some bacteria. *Journal of general microbiology* **64**(2): 187-194.
182. Aulabaugh A, Schloss JV (1990) Oxalyl hydroxamates as reaction-intermediate analogues for ketol-acid reductoisomerase. *Biochemistry* **29**(11): 2824-2830.
183. Durner J, Knorzer OC, Boger P (1993) Ketol-Acid Reductoisomerase from Barley (Hordeum vulgare) (Purification, Properties, and Specific Inhibition). *Plant Physiol* **103**(3): 903-910.
184. Westerfeld WW (1945) A COLORIMETRIC DETERMINATION OF BLOOD ACETOIN. pp. 495-502.
185. Hyduke DR, Jarboe LR, Tran LM, Chou KJ, Liao JC (2007) Integrated network analysis identifies nitric oxide response networks and dihydroxyacid dehydratase as a crucial target in Escherichia coli. *Proceedings of the National Academy of Sciences of the United States of America* **104**(20): 8484-8489.
186. Baba T, Ara T, Hasegawa M, Takai Y, Okumura Y, et al. (2006) Construction of Escherichia coli K-12 in-frame, single-gene knockout mutants: the Keio collection. *Molecular systems biology* **2**: 2006 0008.
187. Joyce AR, Reed JL, White A, Edwards R, Osterman A, et al. (2006) Experimental and computational assessment of conditionally essential genes in Escherichia coli. *Journal of bacteriology* **188**(23): 8259-8271.
188. Massey V, Veeger C (1960) On the reaction mechanism of lipoyl dehydrogenase. *Biochimica et biophysica acta* **40**: 184-185.
189. Li Calzi M, Poole LB (1997) Requirement for the two AhpF cystine disulfide centers in catalysis of peroxide reduction by alkyl hydroperoxide reductase. *Biochemistry* **36**(43): 13357-13364.
190. Liu XI, Korde N, Jakob U, Leichert LI (2006) CoSMoS: Conserved Sequence Motif Search in the proteome. *BMC bioinformatics* **7**: 37.
191. McGuire KA, Siggaard-Andersen M, Bangera MG, Olsen JG, von Wettstein-Knowles P (2001) beta-Ketoacyl-[acyl carrier protein] synthase I of Escherichia coli: aspects of the condensation mechanism revealed by analyses of mutations in the active site pocket. *Biochemistry* **40**(33): 9836-9845.
192. Witkowski A, Joshi AK, Smith S (2002) Mechanism of the beta-ketoacyl synthase reaction catalyzed by the animal fatty acid synthase. *Biochemistry* **41**(35): 10877-10887.
193. Karsten WE, Viola RE (1992) Identification of an essential cysteine in the reaction catalyzed by aspartate-beta-semialdehyde dehydrogenase from Escherichia coli. *Biochimica et biophysica acta* **1121**(1-2): 234-238.
194. Vanoni MA, Verzotti E, Zanetti G, Curti B (1996) Properties of the recombinant beta subunit of glutamate synthase. *European journal of biochemistry / FEBS* **236**(3): 937-946.
195. Anborgh PH, Parmeggiani A, Jonak J (1992) Site-directed mutagenesis of elongation factor Tu. The functional and structural role of residue Cys81. *European Journal of biochemistry / FEBS* **208**(2): 251-257.

196. Hara MR, Snyder SH (2006) Nitric oxide-GAPDH-Siah: a novel cell death cascade. *Cellular and molecular neurobiology* **26**(4-6): 527-538.
197. Jaffrey SR, Erdjument-Bromage H, Ferris CD, Tempst P, Snyder SH (2001) Protein S-nitrosylation: a physiological signal for neuronal nitric oxide. *Nature cell biology* **3**(2): 193-197.
198. Huang B, Chen C (2006) An ascorbate-dependent artifact that interferes with the interpretation of the biotin switch assay. *Free radical biology & medicine* **41**(4): 562-567.
199. Karala AR, Ruddock LW (2007) Does s-methyl methanethiosulfonate trap the thiol-disulfide state of proteins? *Antioxidants & redox signaling* **9**(4): 527-531.
200. Lancaster JR, Jr. (2006) Nitroxidative, nitrosative, and nitrative stress: kinetic predictions of reactive nitrogen species chemistry under biological conditions. *Chemical research in toxicology* **19**(9): 1160-1174.
201. Langley D, Guest JR (1977) Biochemical genetics of the alpha-keto acid dehydrogenase complexes of Escherichia coli K12: isolation and biochemical properties of deletion mutants. *Journal of general microbiology* **99**(2): 263-276.
202. Knowles FC (1985) Reactions of lipoamide dehydrogenase and glutathione reductase with arsonic acids and arsonous acids. *Archives of biochemistry and biophysics* **242**(1): 1-10.
203. de Mendoza D, Klages Ulrich A, Cronan JE, Jr. (1983) Thermal regulation of membrane fluidity in Escherichia coli. Effects of overproduction of beta-ketoacyl-acyl carrier protein synthase I. *The Journal of biological chemistry* **258**(4): 2098-2101.
204. Poole LB (2005) Bacterial defenses against oxidants: mechanistic features of cysteine-based peroxidases and their flavoprotein reductases. *Archives of biochemistry and biophysics* **433**(1): 240-254.
205. Murzin AG, Brenner SE, Hubbard T, Chothia C (1995) SCOP: a structural classification of proteins database for the investigation of sequences and structures. *Journal of molecular biology* **247**(4): 536-540.
206. Sliskovic I, Raturi A, Mutus B (2005) Characterization of the S-denitrosation activity of protein disulfide isomerase. *The Journal of biological chemistry* **280**(10): 8733-8741.
207. Vanoni MA, Edmondson DE, Zanetti G, Curti B (1992) Characterization of the flavins and the iron-sulfur centers of glutamate synthase from Azospirillum brasilense by absorption, circular dichroism, and electron paramagnetic resonance spectroscopies. *Biochemistry* **31**(19): 4613-4623.
208. Ding H, Demple B (2000) Direct nitric oxide signal transduction via nitrosylation of iron-sulfur centers in the SoxR transcription activator. *Proceedings of the National Academy of Sciences of the United States of America* **97**(10): 5146-5150.
209. Hondorp ER, Matthews RG (2004) Oxidative stress inactivates cobalamin-independent methionine synthase (MetE) in Escherichia coli. *PLoS biology* **2**(11): e336.
210. Tempest DW, Meers JL, Brown CM (1970) Synthesis of glutamate in Aerobacter aerogenes by a hitherto unknown route. *The Biochemical journal* **117**(2): 405-407.
211. Reitzer L (2003) Nitrogen assimilation and global regulation in Escherichia coli. *Annual review of microbiology* **57**: 155-176.
212. Miller RE, Stadtman ER (1972) Glutamate synthase from Escherichia coli. An iron-sulfide flavoprotein. *The Journal of biological chemistry* **247**(22): 7407-7419.
213. Agnelli P, Dossena L, Colombi P, Mulazzi S, Morandi P, et al. (2005) The unexpected structural role of glutamate synthase [4Fe-4S](+1,+2) clusters as demonstrated by site-directed mutagenesis of conserved C residues at the N-terminus of the enzyme beta subunit. *Archives of biochemistry and biophysics* **436**(2): 355-366.
214. Lancaster JR, Jr. (1997) A tutorial on the diffusibility and reactivity of free nitric oxide. *Nitric Oxide* **1**(1): 18-30.

215. De Groote MA, Granger D, Xu Y, Campbell G, Prince R, et al. (1995) Genetic and redox determinants of nitric oxide cytotoxicity in a Salmonella typhimurium model. *Proceedings of the National Academy of Sciences of the United States of America* **92**(14): 6399-6403.
216. Spiro S (2006) Nitric oxide-sensing mechanisms in Escherichia coli. *Biochemical Society transactions* **34**(Pt 1): 200-202.
217. Stamler JS, Simon DI, Osborne JA, Mullins ME, Jaraki O, et al. (1992) S-nitrosylation of proteins with nitric oxide: synthesis and characterization of biologically active compounds. *Proceedings of the National Academy of Sciences of the United States of America* **89**(1): 444-448.
218. Orme-Johnson WH, Orme-Johnson NR (1978) Overview of iron--sulfur proteins. *Methods in enzymology* **53**: 259-268.
219. Hernandez-Fonseca K, Massieu L (2005) Disruption of endoplasmic reticulum calcium stores is involved in neuronal death induced by glycolysis inhibition in cultured hippocampal neurons. *Journal of neuroscience research* **82**(2): 196-205.
220. Vanoni MA, Curti B (1999) Glutamate synthase: a complex iron-sulfur flavoprotein. *Cell Mol Life Sci* **55**(4): 617-638.
221. Gerdes SY, Scholle MD, Campbell JW, Balazsi G, Ravasz E, et al. (2003) Experimental determination and system level analysis of essential genes in Escherichia coli MG1655. *Journal of bacteriology* **185**(19): 5673-5684.
222. Kroncke KD, Klotz LO, Suschek CV, Sies H (2002) Comparing nitrosative versus oxidative stress toward zinc finger-dependent transcription. Unique role for NO. *The Journal of biological chemistry* **277**(15): 13294-13301.
223. Klatt P, Lamas S (2000) Regulation of protein function by S-glutathiolation in response to oxidative and nitrosative stress. *European journal of biochemistry / FEBS* **267**(16): 4928-4944.
224. Ascenzi P, Colasanti M, Persichini T, Muolo M, Polticelli F, et al. (2000) Re-evaluation of amino acid sequence and structural consensus rules for cysteine-nitric oxide reactivity. *Biological chemistry* **381**(7): 623-627.
225. Winter J, Ilbert M, Graf PC, Ozcelik D, Jakob U (2008) Bleach activates a redox-regulated chaperone by oxidative protein unfolding. *Cell* **135**(4): 691-701.
226. Suliman HB, Carraway MS, Piantadosi CA (2003) Postlipopolysaccharide oxidative damage of mitochondrial DNA. *American journal of respiratory and critical care medicine* **167**(4): 570-579.
227. Wei YH, Lu CY, Lee HC, Pang CY, Ma YS (1998) Oxidative damage and mutation to mitochondrial DNA and age-dependent decline of mitochondrial respiratory function. *Annals of the New York Academy of Sciences* **854**: 155-170.
228. Choi TY, Park SY, Kang HS, Cheong JH, Kim HD, et al. (2004) Redox regulation of DNA binding activity of DREF (DNA replication-related element binding factor) in Drosophila. *The Biochemical journal* **378**(Pt 3): 833-838.
229. Turrens JF (1997) Superoxide production by the mitochondrial respiratory chain. *Bioscience reports* **17**(1): 3-8.
230. Droge W (2002) Free radicals in the physiological control of cell function. *Physiological reviews* **82**(1): 47-95.
231. Cohen G (2000) Oxidative stress, mitochondrial respiration, and Parkinson's disease. *Annals of the New York Academy of Sciences* **899**: 112-120.
232. Lodi R, Tonon C, Calabrese V, Schapira AH (2006) Friedreich's ataxia: from disease mechanisms to therapeutic interventions. *Antioxidants & redox signaling* **8**(3-4): 438-443.
233. McGill JK, Beal MF (2006) PGC-1alpha, a new therapeutic target in Huntington's disease? *Cell* **127**(3): 465-468.
234. Donath MY, Ehses JA, Maedler K, Schumann DM, Ellingsgaard H, et al. (2005) Mechanisms of beta-cell death in type 2 diabetes. *Diabetes* **54 Suppl 2**: S108-113.

235. Zhang P, Davis AT, Ahmed K (1998) Mechanism of protein kinase CK2 association with nuclear matrix: role of disulfide bond formation. *Journal of cellular biochemistry* **69**(2): 211-220.
236. Dixon BM, Heath SH, Kim R, Suh JH, Hagen TM (2008) Assessment of endoplasmic reticulum glutathione redox status is confounded by extensive ex vivo oxidation. *Antioxidants & redox signaling* **10**(5): 963-972.
237. Halvey PJ, Watson WH, Hansen JM, Go YM, Samali A, et al. (2005) Compartmental oxidation of thiol-disulphide redox couples during epidermal growth factor signalling. *The Biochemical journal* **386**(Pt 2): 215-219.
238. Nkabyo YS, Ziegler TR, Gu LH, Watson WH, Jones DP (2002) Glutathione and thioredoxin redox during differentiation in human colon epithelial (Caco-2) cells. *American journal of physiology* **283**(6): G1352-1359.
239. Ostergaard H, Tachibana C, Winther JR (2004) Monitoring disulfide bond formation in the eukaryotic cytosol. *The Journal of cell biology* **166**(3): 337-345.
240. Hanson GT, Aggeler R, Oglesbee D, Cannon M, Capaldi RA, et al. (2004) Investigating mitochondrial redox potential with redox-sensitive green fluorescent protein indicators. *The Journal of biological chemistry* **279**(13): 13044-13053.
241. Shen D, Dalton TP, Nebert DW, Shertzer HG (2005) Glutathione redox state regulates mitochondrial reactive oxygen production. *The Journal of biological chemistry* **280**(27): 25305-25312.
242. Hwang C, Sinskey AJ, Lodish HF (1992) Oxidized redox state of glutathione in the endoplasmic reticulum. *Science (New York, NY* **257**(5076): 1496-1502.
243. Hansen JM, Go YM, Jones DP (2006) Nuclear and mitochondrial compartmentation of oxidative stress and redox signaling. *Annual review of pharmacology and toxicology* **46**: 215-234.
244. Lopez-Mirabal HR, Winther JR (2008) Redox characteristics of the eukaryotic cytosol. *Biochimica et biophysica acta* **1783**(4): 629-640.
245. Le Moan N, Clement G, Le Maout S, Tacnet F, Toledano MB (2006) The Saccharomyces cerevisiae proteome of oxidized protein thiols: contrasted functions for the thioredoxin and glutathione pathways. *The Journal of biological chemistry* **281**(15): 10420-10430.
246. Fratelli M, Demol H, Puype M, Casagrande S, Eberini I, et al. (2002) Identification by redox proteomics of glutathionylated proteins in oxidatively stressed human T lymphocytes. *Proceedings of the National Academy of Sciences of the United States of America* **99**(6): 3505-3510.
247. Bondar RJ, Mead DC (1974) Evaluation of glucose-6-phosphate dehydrogenase from Leuconostoc mesenteroides in the hexokinase method for determining glucose in serum. *Clinical chemistry* **20**(5): 586-590.
248. Meisinger C, Sommer T, Pfanner N (2000) Purification of Saccharomcyes cerevisiae mitochondria devoid of microsomal and cytosolic contaminations. *Analytical biochemistry* **287**(2): 339-342.
249. Haas A (1995) A quantitative assay to measure homotypic vacuole fusion in vitro. *Methods in Cell Science* **17**(4): 283-294.
250. Dove JE, Brockenbrough JS, Aris JP (1998) Isolation of nuclei and nucleoli from the yeast Saccharomyces cerevisiae. *Methods in cell biology* **53**: 33-46.
251. Bellew M, Coram M, Fitzgibbon M, Igra M, Randolph T, et al. (2006) A suite of algorithms for the comprehensive analysis of complex protein mixtures using high-resolution LC-MS. *Bioinformatics (Oxford, England)* **22**(15): 1902-1909.
252. Li H, Robertson AD, Jensen JH (2005) Very fast empirical prediction and rationalization of protein pKa values. *Proteins* **61**(4): 704-721.
253. Frishman D, Argos P (1995) Knowledge-based protein secondary structure assignment. *Proteins* **23**(4): 566-579.

254. Prilusky J, Felder CE, Zeev-Ben-Mordehai T, Rydberg EH, Man O, et al. (2005) FoldIndex: a simple tool to predict whether a given protein sequence is intrinsically unfolded. *Bioinformatics (Oxford, England)* **21**(16): 3435-3438.
255. Sarry JE, Chen S, Collum RP, Liang S, Peng M, et al. (2007) Analysis of the vacuolar luminal proteome of Saccharomyces cerevisiae. *The FEBS journal* **274**(16): 4287-4305.
256. Sickmann A, Reinders J, Wagner Y, Joppich C, Zahedi R, et al. (2003) The proteome of Saccharomyces cerevisiae mitochondria. *Proceedings of the National Academy of Sciences of the United States of America* **100**(23): 13207-13212.
257. McDonagh B, Ogueta S, Lasarte G, Padilla CA, Barcena JA (2009) Shotgun redox proteomics identifies specifically modified cysteines in key metabolic enzymes under oxidative stress in Saccharomyces cerevisiae. *Journal of proteomics* **72**(4): 677-689.
258. Gygi SP, Rist B, Gerber SA, Turecek F, Gelb MH, et al. (1999) Quantitative analysis of complex protein mixtures using isotope-coded affinity tags. *Nature biotechnology* **17**(10): 994-999.
259. Kozarova A, Sliskovic I, Mutus B, Simon ES, Andrews PC, et al. (2007) Identification of redox sensitive thiols of protein disulfide isomerase using isotope coded affinity technology and mass spectrometry. *Journal of the American Society for Mass Spectrometry* **18**(2): 260-269.
260. Sethuraman M, Clavreul N, Huang H, McComb ME, Costello CE, et al. (2007) Quantification of oxidative posttranslational modifications of cysteine thiols of p21ras associated with redox modulation of activity using isotope-coded affinity tags and mass spectrometry. *Free radical biology & medicine* **42**(6): 823-829.
261. Le Moan N, Tacnet F, Toledano MB (2009) Protein-thiol oxidation, from single proteins to proteome-wide analyses. *Methods in molecular biology (Clifton, NJ* **476**: 175-192.
262. Cabiscol E, Piulats E, Echave P, Herrero E, Ros J (2000) Oxidative stress promotes specific protein damage in Saccharomyces cerevisiae. *The Journal of biological chemistry* **275**(35): 27393-27398.
263. Magherini F, Carpentieri A, Amoresano A, Gamberi T, De Filippo C, et al. (2009) Different carbon sources affect lifespan and protein redox state during Saccharomyces cerevisiae chronological ageing. *Cell Mol Life Sci* **66**(5): 933-947.
264. Hu J, Dong L, Outten CE (2008) The redox environment in the mitochondrial intermembrane space is maintained separately from the cytosol and matrix. *The Journal of biological chemistry* **283**(43): 29126-29134.
265. Lewis JG, Northcott CJ, Learmonth RP, Attfield PV, Watson K (1993) The need for consistent nomenclature and assessment of growth phases in diauxic cultures of Saccharomyces cerevisiae. *Journal of general microbiology* **139**(4): 835-839.
266. O'Brien KM, Dirmeier R, Engle M, Poyton RO (2004) Mitochondrial protein oxidation in yeast mutants lacking manganese-(MnSOD) or copper- and zinc-containing superoxide dismutase (CuZnSOD): evidence that MnSOD and CuZnSOD have both unique and overlapping functions in protecting mitochondrial proteins from oxidative damage. *The Journal of biological chemistry* **279**(50): 51817-51827.
267. Kitanovic A, Walther T, Loret MO, Holzwarth J, Kitanovic I, et al. (2009) Metabolic response to MMS-mediated DNA damage in Saccharomyces cerevisiae is dependent on the glucose concentration in the medium. *FEMS yeast research* **9**(4): 535-551.
268. Fabrizio P, Longo VD (2003) The chronological life span of Saccharomyces cerevisiae. *Aging cell* **2**(2): 73-81.
269. Toledano MB, Kumar C, Le Moan N, Spector D, Tacnet F (2007) The system biology of thiol redox system in Escherichia coli and yeast: differential functions in oxidative stress, iron metabolism and DNA synthesis. *FEBS letters* **581**(19): 3598-3607.

270. Furukawa Y, Torres AS, O'Halloran TV (2004) Oxygen-induced maturation of SOD1: a key role for disulfide formation by the copper chaperone CCS. *The EMBO journal* **23**(14): 2872-2881.
271. Allen S, Lu H, Thornton D, Tokatlidis K (2003) Juxtaposition of the two distal CX3C motifs via intrachain disulfide bonding is essential for the folding of Tim10. *The Journal of biological chemistry* **278**(40): 38505-38513.
272. Fridovich I (1986) Superoxide dismutases. *Advances in enzymology and related areas of molecular biology* **58**: 61-97.
273. Crapo JD, Oury T, Rabouille C, Slot JW, Chang LY (1992) Copper,zinc superoxide dismutase is primarily a cytosolic protein in human cells. *Proceedings of the National Academy of Sciences of the United States of America* **89**(21): 10405-10409.
274. Hornberg A, Logan DT, Marklund SL, Oliveberg M (2007) The coupling between disulphide status, metallation and dimer interface strength in Cu/Zn superoxide dismutase. *Journal of molecular biology* **365**(2): 333-342.
275. Leitch JM, Jensen LT, Bouldin SD, Outten CE, Hart PJ, et al. (2009) Activation of Cu,Zn-superoxide dismutase in the absence of oxygen and the copper chaperone CCS. *The Journal of biological chemistry* **284**(33): 21863-21871.
276. Arnesano F, Banci L, Bertini I, Martinelli M, Furukawa Y, et al. (2004) The unusually stable quaternary structure of human Cu,Zn-superoxide dismutase 1 is controlled by both metal occupancy and disulfide status. *The Journal of biological chemistry* **279**(46): 47998-48003.
277. Field LS, Furukawa Y, O'Halloran TV, Culotta VC (2003) Factors controlling the uptake of yeast copper/zinc superoxide dismutase into mitochondria. *The Journal of biological chemistry* **278**(30): 28052-28059.
278. Hatfield PM, Vierstra RD (1992) Multiple forms of ubiquitin-activating enzyme E1 from wheat. Identification of an essential cysteine by in vitro mutagenesis. *The Journal of biological chemistry* **267**(21): 14799-14803.
279. Costa VM, Amorim MA, Quintanilha A, Moradas-Ferreira P (2002) Hydrogen peroxide-induced carbonylation of key metabolic enzymes in Saccharomyces cerevisiae: the involvement of the oxidative stress response regulators Yap1 and Skn7. *Free radical biology & medicine* **33**(11): 1507-1515.
280. Tamarit J, Cabiscol E, Ros J (1998) Identification of the major oxidatively damaged proteins in Escherichia coli cells exposed to oxidative stress. *The Journal of biological chemistry* **273**(5): 3027-3032.
281. Reverter-Branchat G, Cabiscol E, Tamarit J, Ros J (2004) Oxidative damage to specific proteins in replicative and chronological-aged Saccharomyces cerevisiae: common targets and prevention by calorie restriction. *The Journal of biological chemistry* **279**(30): 31983-31989.
282. Magherini F, Tani C, Gamberi T, Caselli A, Bianchi L, et al. (2007) Protein expression profiles in Saccharomyces cerevisiae during apoptosis induced by H2O2. *Proteomics* **7**(9): 1434-1445.
283. Baty JW, Hampton MB, Winterbourn CC (2005) Proteomic detection of hydrogen peroxide-sensitive thiol proteins in Jurkat cells. *The Biochemical journal* **389**(Pt 3): 785-795.
284. Saurin AT, Neubert H, Brennan JP, Eaton P (2004) Widespread sulfenic acid formation in tissues in response to hydrogen peroxide. *Proceedings of the National Academy of Sciences of the United States of America* **101**(52): 17982-17987.
285. Giaever G, Chu AM, Ni L, Connelly C, Riles L, et al. (2002) Functional profiling of the Saccharomyces cerevisiae genome. *Nature* **418**(6896): 387-391.
286. Grant CM, Quinn KA, Dawes IW (1999) Differential protein S-thiolation of glyceraldehyde-3-phosphate dehydrogenase isoenzymes influences sensitivity to oxidative stress. *Molecular and cellular biology* **19**(4): 2650-2656.
287. Mattiasson B, Hahn-Hägerdal B (1982) Microenvironmental effects on metabolic behaviour of immobilized cells a hypothesis. *Applied Microbiology and Biotechnology* **16**(1): 52-55.

288. den Hollander JA, Ugurbil K, Brown TR, Bednar M, Redfield C, et al. (1986) Studies of anaerobic and aerobic glycolysis in Saccharomyces cerevisiae. *Biochemistry* **25**(1): 203-211.
289. Lagunas R, Gancedo C (1983) Role of phosphate in the regulation of the Pasteur effect in Saccharomyces cerevisiae. *European journal of biochemistry / FEBS* **137**(3): 479-483.
290. Miles EW, Higgins W (1980) Location of the reactive sulfhydryl residues in the primary sequence of the beta 2 subunit of tryptophan synthase of Escherichia coli. *Biochemical and biophysical research communications* **93**(4): 1152-1159.
291. Miles EW, Kawasaki H, Ahmed SA, Morita H, Morita H, et al. (1989) The beta subunit of tryptophan synthase. Clarification of the roles of histidine 86, lysine 87, arginine 148, cysteine 170, and cysteine 230. *The Journal of biological chemistry* **264**(11): 6280-6287.
292. Zalkin H, Yanofsky C (1982) Yeast gene TRP5: structure, function, regulation. *The Journal of biological chemistry* **257**(3): 1491-1500.
293. Helmstaedt K, Strittmatter A, Lipscomb WN, Braus GH (2005) Evolution of 3-deoxy-D-arabino-heptulosonate-7-phosphate synthase-encoding genes in the yeast Saccharomyces cerevisiae. *Proceedings of the National Academy of Sciences of the United States of America* **102**(28): 9784-9789.
294. Lin LL, Liao HF, Chien HR, Hsu WH (2001) Identification of essential cysteine residues in 3-deoxy-D-arabino-heptulosonate-7-phosphate synthase from Corynebacterium glutamicum. *Current microbiology* **42**(6): 426-431.
295. Dever TE (2002) Gene-specific regulation by general translation factors. *Cell* **108**(4): 545-556.
296. Dunand-Sauthier I, Walker CA, Narasimhan J, Pearce AK, Wek RC, et al. (2005) Stress-activated protein kinase pathway functions to support protein synthesis and translational adaptation in response to environmental stress in fission yeast. *Eukaryotic cell* **4**(11): 1785-1793.
297. Kumsta C, Jakob U (2009) Redox-Regulated Chaperones. *Biochemistry*.
298. Trivelli X, Krimm I, Ebel C, Verdoucq L, Prouzet-Mauleon V, et al. (2003) Characterization of the yeast peroxiredoxin Ahp1 in its reduced active and overoxidized inactive forms using NMR. *Biochemistry* **42**(48): 14139-14149.
299. Vignols F, Mouaheb N, Thomas D, Meyer Y (2003) Redox control of Hsp70-Co-chaperone interaction revealed by expression of a thioredoxin-like Arabidopsis protein. *The Journal of biological chemistry* **278**(7): 4516-4523.
300. Mirzaei H, Regnier F (2006) Creation of allotypic active sites during oxidative stress. *Journal of proteome research* **5**(9): 2159-2168.
301. Chivers PT, Prehoda KE, Volkman BF, Kim BM, Markley JL, et al. (1997) Microscopic pKa values of Escherichia coli thioredoxin. *Biochemistry* **36**(48): 14985-14991.
302. Cocheme HM, Murphy MP (2008) Complex I is the major site of mitochondrial superoxide production by paraquat. *The Journal of biological chemistry* **283**(4): 1786-1798.
303. Hochgrafe F, Mostertz J, Albrecht D, Hecker M (2005) Fluorescence thiol modification assay: oxidatively modified proteins in Bacillus subtilis. *Molecular microbiology* **58**(2): 409-425.
304. Gardner PR, Fridovich I (1991) Superoxide sensitivity of the Escherichia coli aconitase. *The Journal of biological chemistry* **266**(29): 19328-19333.
305. Gardner PR, Fridovich I (1991) Superoxide sensitivity of the Escherichia coli 6-phosphogluconate dehydratase. *The Journal of biological chemistry* **266**(3): 1478-1483.
306. Flint DH, Tuminello JF, Emptage MH (1993) The inactivation of Fe-S cluster containing hydro-lyases by superoxide. *The Journal of biological chemistry* **268**(30): 22369-22376.
307. Liochev SI, Fridovich I (1994) The role of O2.- in the production of HO.: in vitro and in vivo. *Free radical biology & medicine* **16**(1): 29-33.
308. Keyer K, Imlay JA (1996) Superoxide accelerates DNA damage by elevating free-iron levels. *Proceedings of the National Academy of Sciences of the United States of America* **93**(24): 13635-13640.

309. Sinha A, Maitra PK (1992) Induction of specific enzymes of the oxidative pentose phosphate pathway by glucono-delta-lactone in Saccharomyces cerevisiae. *Journal of general microbiology* **138**(9): 1865-1873.
310. Nogae I, Johnston M (1990) Isolation and characterization of the ZWF1 gene of Saccharomyces cerevisiae, encoding glucose-6-phosphate dehydrogenase. *Gene* **96**(2): 161-169.
311. Bus JS, Gibson JE (1984) Paraquat: model for oxidant-initiated toxicity. *Environmental health perspectives* **55**: 37-46.
312. Longo VD, Liou LL, Valentine JS, Gralla EB (1999) Mitochondrial superoxide decreases yeast survival in stationary phase. *Archives of biochemistry and biophysics* **365**(1): 131-142.
313. Flint DH, Tuminello JF, Miller TJ (1996) Studies on the synthesis of the Fe-S cluster of dihydroxy-acid dehydratase in escherichia coli crude extract. Isolation of O-acetylserine sulfhydrylases A and B and beta-cystathionase based on their ability to mobilize sulfur from cysteine and to participate in Fe-S cluster synthesis. *The Journal of biological chemistry* **271**(27): 16053-16067.
314. Brandes N, Schmitt S, Jakob U (2008) Thiol-Based Redox Switches in Eukaryotic Proteins. *Antioxidants & redox signaling*.
315. Beal MF (2002) Oxidatively modified proteins in aging and disease. *Free radical biology & medicine* **32**(9): 797-803.
316. Butterfield DA, Lauderback CM (2002) Lipid peroxidation and protein oxidation in Alzheimer's disease brain: potential causes and consequences involving amyloid beta-peptide-associated free radical oxidative stress. *Free radical biology & medicine* **32**(11): 1050-1060.
317. Ischiropoulos H, Beckman JS (2003) Oxidative stress and nitration in neurodegeneration: cause, effect, or association? *The Journal of clinical investigation* **111**(2): 163-169.
318. Thomas JA, Mallis RJ (2001) Aging and oxidation of reactive protein sulfhydryls. *Experimental gerontology* **36**(9): 1519-1526.
319. Finkel E (2002) Disulfide bond switches. *Nature biotechnology* **20**(9): 887.
320. Souza JM, Radi R (1998) Glyceraldehyde-3-phosphate dehydrogenase inactivation by peroxynitrite. *Archives of biochemistry and biophysics* **360**(2): 187-194.
321. Zheng M, Aslund F, Storz G (1998) Activation of the OxyR transcription factor by reversible disulfide bond formation. *Science (New York, NY* **279**(5357): 1718-1721.
322. Nystrom T (2002) Translational fidelity, protein oxidation, and senescence: lessons from bacteria. *Ageing research reviews* **1**(4): 693-703.
323. Hansen RE, Roth D, Winther JR (2009) Quantifying the global cellular thiol-disulfide status. *Proceedings of the National Academy of Sciences of the United States of America* **106**(2): 422-427.
324. Davies MJ (2005) The oxidative environment and protein damage. *Biochimica et biophysica acta* **1703**(2): 93-109.
325. Stadtman ER, Levine RL (2003) Free radical-mediated oxidation of free amino acids and amino acid residues in proteins. *Amino acids* **25**(3-4): 207-218.
326. Lind C, Gerdes R, Hamnell Y, Schuppe-Koistinen I, von Lowenhielm HB, et al. (2002) Identification of S-glutathionylated cellular proteins during oxidative stress and constitutive metabolism by affinity purification and proteomic analysis. *Archives of biochemistry and biophysics* **406**(2): 229-240.
327. Brennan JP, Wait R, Begum S, Bell JR, Dunn MJ, et al. (2004) Detection and mapping of widespread intermolecular protein disulfide formation during cardiac oxidative stress using proteomics with diagonal electrophoresis. *The Journal of biological chemistry* **279**(40): 41352-41360.
328. Minard KI, Carroll CA, Weintraub ST, Mc-Alister-Henn L (2007) Changes in disulfide bond content of proteins in a yeast strain lacking major sources of NADPH. *Free radical biology & medicine* **42**(1): 106-117.

329. Yoo BS, Regnier FE (2004) Proteomic analysis of carbonylated proteins in two-dimensional gel electrophoresis using avidin-fluorescein affinity staining. *Electrophoresis* **25**(9): 1334-1341.
330. Harman D (1956) Aging: a theory based on free radical and radiation chemistry. *Journal of gerontology* **11**(3): 298-300.
331. Beckman KB, Ames BN (1998) The free radical theory of aging matures. *Physiological reviews* **78**(2): 547-581.
332. Bitterman KJ, Medvedik O, Sinclair DA (2003) Longevity regulation in Saccharomyces cerevisiae: linking metabolism, genome stability, and heterochromatin. *Microbiol Mol Biol Rev* **67**(3): 376-399, table of contents.
333. Longo VD (1999) Mutations in signal transduction proteins increase stress resistance and longevity in yeast, nematodes, fruit flies, and mammalian neuronal cells. *Neurobiology of aging* **20**(5): 479-486.
334. MacLean M, Harris N, Piper PW (2001) Chronological lifespan of stationary phase yeast cells; a model for investigating the factors that might influence the ageing of postmitotic tissues in higher organisms. *Yeast (Chichester, England)* **18**(6): 499-509.
335. Grant CM, Perrone G, Dawes IW (1998) Glutathione and catalase provide overlapping defenses for protection against hydrogen peroxide in the yeast Saccharomyces cerevisiae. *Biochemical and biophysical research communications* **253**(3): 893-898.
336. Berlett BS, Stadtman ER (1997) Protein oxidation in aging, disease, and oxidative stress. *The Journal of biological chemistry* **272**(33): 20313-20316.
337. Smith CD, Carney JM, Starke-Reed PE, Oliver CN, Stadtman ER, et al. (1991) Excess brain protein oxidation and enzyme dysfunction in normal aging and in Alzheimer disease. *Proceedings of the National Academy of Sciences of the United States of America* **88**(23): 10540-10543.
338. Chen Q, Ding Q, Keller JN (2005) The stationary phase model of aging in yeast for the study of oxidative stress and age-related neurodegeneration. *Biogerontology* **6**(1): 1-13.
339. Harris N, Bachler M, Costa V, Mollapour M, Moradas-Ferreira P, et al. (2005) Overexpressed Sod1p acts either to reduce or to increase the lifespans and stress resistance of yeast, depending on whether it is Cu(2+)-deficient or an active Cu,Zn-superoxide dismutase. *Aging cell* **4**(1): 41-52.
340. Fabrizio P, Liou LL, Moy VN, Diaspro A, Valentine JS, et al. (2003) SOD2 functions downstream of Sch9 to extend longevity in yeast. *Genetics* **163**(1): 35-46.
341. Kaeberlein M, Burtner CR, Kennedy BK (2007) Recent developments in yeast aging. *PLoS genetics* **3**(5): e84.
342. Fabrizio P, Longo VD (2007) The chronological life span of Saccharomyces cerevisiae. *Methods in molecular biology (Clifton, NJ* **371**: 89-95.
343. Longo VD, Ellerby LM, Bredesen DE, Valentine JS, Gralla EB (1997) Human Bcl-2 reverses survival defects in yeast lacking superoxide dismutase and delays death of wild-type yeast. *The Journal of cell biology* **137**(7): 1581-1588.
344. Longo VD, Gralla EB, Valentine JS (1996) Superoxide dismutase activity is essential for stationary phase survival in Saccharomyces cerevisiae. Mitochondrial production of toxic oxygen species in vivo. *The Journal of biological chemistry* **271**(21): 12275-12280.
345. Pocsi I, Prade RA, Penninckx MJ (2004) Glutathione, altruistic metabolite in fungi. *Advances in microbial physiology* **49**: 1-76.
346. Harris N, Costa V, MacLean M, Mollapour M, Moradas-Ferreira P, et al. (2003) Mnsod overexpression extends the yeast chronological (G(0)) life span but acts independently of Sir2p histone deacetylase to shorten the replicative life span of dividing cells. *Free radical biology & medicine* **34**(12): 1599-1606.
347. Herker E, Jungwirth H, Lehmann KA, Maldener C, Frohlich KU, et al. (2004) Chronological aging leads to apoptosis in yeast. *The Journal of cell biology* **164**(4): 501-507.

348. Masoro EJ (2005) Overview of caloric restriction and ageing. *Mechanisms of ageing and development* **126**(9): 913-922.
349. Sinclair DA (2005) Toward a unified theory of caloric restriction and longevity regulation. *Mechanisms of ageing and development* **126**(9): 987-1002.
350. Mair W, Dillin A (2008) Aging and survival: the genetics of life span extension by dietary restriction. *Annual review of biochemistry* **77**: 727-754.
351. Lin SJ, Kaeberlein M, Andalis AA, Sturtz LA, Defossez PA, et al. (2002) Calorie restriction extends Saccharomyces cerevisiae lifespan by increasing respiration. *Nature* **418**(6895): 344-348.
352. Murakami CJ, Burtner CR, Kennedy BK, Kaeberlein M (2008) A method for high-throughput quantitative analysis of yeast chronological life span. *The journals of gerontology* **63**(2): 113-121.
353. Smith DL, Jr., McClure JM, Matecic M, Smith JS (2007) Calorie restriction extends the chronological lifespan of Saccharomyces cerevisiae independently of the Sirtuins. *Aging cell* **6**(5): 649-662.
354. Bonawitz ND, Chatenay-Lapointe M, Pan Y, Shadel GS (2007) Reduced TOR signaling extends chronological life span via increased respiration and upregulation of mitochondrial gene expression. *Cell metabolism* **5**(4): 265-277.
355. Fabrizio P, Pletcher SD, Minois N, Vaupel JW, Longo VD (2004) Chronological aging-independent replicative life span regulation by Msn2/Msn4 and Sod2 in Saccharomyces cerevisiae. *FEBS letters* **557**(1-3): 136-142.
356. Bouillaud F, Arechaga I, Petit PX, Raimbault S, Levi-Meyrueis C, et al. (1994) A sequence related to a DNA recognition element is essential for the inhibition by nucleotides of proton transport through the mitochondrial uncoupling protein. *The EMBO journal* **13**(8): 1990-1997.
357. Saeed AI, Bhagabati NK, Braisted JC, Liang W, Sharov V, et al. (2006) TM4 microarray software suite. *Methods in enzymology* **411**: 134-193.
358. Getz G, Levine E, Domany E (2000) Coupled two-way clustering analysis of gene microarray data. *Proceedings of the National Academy of Sciences of the United States of America* **97**(22): 12079-12084.
359. Yang NC, Ho WM, Chen YH, Hu ML (2002) A convenient one-step extraction of cellular ATP using boiling water for the luciferin-luciferase assay of ATP. *Analytical biochemistry* **306**(2): 323-327.
360. Herbst R, Schafer U, Seckler R (1997) Equilibrium intermediates in the reversible unfolding of firefly (Photinus pyralis) luciferase. *The Journal of biological chemistry* **272**(11): 7099-7105.
361. Hwang C, Lodish HF, Sinskey AJ (1995) Measurement of glutathione redox state in cytosol and secretory pathway of cultured cells. *Methods in enzymology* **251**: 212-221.
362. Kaeberlein M, Andalis AA, Fink GR, Guarente L (2002) High osmolarity extends life span in Saccharomyces cerevisiae by a mechanism related to calorie restriction. *Molecular and cellular biology* **22**(22): 8056-8066.
363. Burtner CR, Murakami CJ, Kennedy BK, Kaeberlein M (2009) A molecular mechanism of chronological aging in yeast. *Cell cycle (Georgetown, Tex* **8**(8): 1256-1270.
364. Fabrizio P, Battistella L, Vardavas R, Gattazzo C, Liou LL, et al. (2004) Superoxide is a mediator of an altruistic aging program in Saccharomyces cerevisiae. *The Journal of cell biology* **166**(7): 1055-1067.
365. Galiazzo F, Labbe-Bois R (1993) Regulation of Cu,Zn- and Mn-superoxide dismutase transcription in Saccharomyces cerevisiae. *FEBS letters* **315**(2): 197-200.
366. Gancedo JM (1992) Carbon catabolite repression in yeast. *European journal of biochemistry / FEBS* **206**(2): 297-313.

367. Lushchak V, Semchyshyn H, Mandryk S, Lushchak O (2005) Possible role of superoxide dismutases in the yeast Saccharomyces cerevisiae under respiratory conditions. *Archives of biochemistry and biophysics* **441**(1): 35-40.
368. Perrone GG, Tan SX, Dawes IW (2008) Reactive oxygen species and yeast apoptosis. *Biochimica et biophysica acta* **1783**(7): 1354-1368.
369. Padgett CM, Whorton AR (1995) S-nitrosoglutathione reversibly inhibits GAPDH by S-nitrosylation. *The American journal of physiology* **269**(3 Pt 1): C739-749.
370. Berlett BS, Levine RL, Stadtman ER (1996) Comparison of the effects of ozone on the modification of amino acid residues in glutamine synthetase and bovine serum albumin. *The Journal of biological chemistry* **271**(8): 4177-4182.
371. Trotter EW, Grant CM (2005) Overlapping roles of the cytoplasmic and mitochondrial redox regulatory systems in the yeast Saccharomyces cerevisiae. *Eukaryotic cell* **4**(2): 392-400.
372. Mustacich D, Powis G (2000) Thioredoxin reductase. *The Biochemical journal* **346 Pt 1**: 1-8.
373. Ghaemmaghami S, Huh WK, Bower K, Howson RW, Belle A, et al. (2003) Global analysis of protein expression in yeast. *Nature* **425**(6959): 737-741.
374. Minard KI, McAlister-Henn L (2001) Antioxidant function of cytosolic sources of NADPH in yeast. *Free radical biology & medicine* **31**(6): 832-843.
375. DeRisi JL, Iyer VR, Brown PO (1997) Exploring the metabolic and genetic control of gene expression on a genomic scale. *Science (New York, NY* **278**(5338): 680-686.
376. Contreras-Shannon V, McAlister-Henn L (2004) Influence of compartmental localization on the function of yeast NADP+-specific isocitrate dehydrogenases. *Archives of biochemistry and biophysics* **423**(2): 235-246.
377. Muller JM, Meyer HH, Ruhrberg C, Stamp GW, Warren G, et al. (1999) The mouse p97 (CDC48) gene. Genomic structure, definition of transcriptional regulatory sequences, gene expression, and characterization of a pseudogene. *The Journal of biological chemistry* **274**(15): 10154-10162.
378. Frohlich KU, Fries HW, Rudiger M, Erdmann R, Botstein D, et al. (1991) Yeast cell cycle protein CDC48p shows full-length homology to the mammalian protein VCP and is a member of a protein family involved in secretion, peroxisome formation, and gene expression. *The Journal of cell biology* **114**(3): 443-453.
379. Noguchi M, Takata T, Kimura Y, Manno A, Murakami K, et al. (2005) ATPase activity of p97/valosin-containing protein is regulated by oxidative modification of the evolutionally conserved cysteine 522 residue in Walker A motif. *The Journal of biological chemistry* **280**(50): 41332-41341.
380. Zabriskie TM, Jackson MD (2000) Lysine biosynthesis and metabolism in fungi. *Natural product reports* **17**(1): 85-97.
381. Barritt (1985) in Pyruvate Carboxylase (Keech, D. B. and Wallace, J. C. eds.), pp. 141±177, CRC Series in Enzyme Biology, CRC Press, Boca Raton, FL.
382. Levert KL, Lloyd RB, Waldrop GL (2000) Do cysteine 230 and lysine 238 of biotin carboxylase play a role in the activation of biotin? *Biochemistry* **39**(14): 4122-4128.
383. Lewis VA, Hynes GM, Zheng D, Saibil H, Willison K (1992) T-complex polypeptide-1 is a subunit of a heteromeric particle in the eukaryotic cytosol. *Nature* **358**(6383): 249-252.
384. Ellis RJ (1990) Molecular Chaperones: The Plant Connection. *Science (New York, NY* **250**(4983): 954-959.
385. Frydman J, Nimmesgern E, Erdjument-Bromage H, Wall JS, Tempst P, et al. (1992) Function in protein folding of TRiC, a cytosolic ring complex containing TCP-1 and structurally related subunits. *The EMBO journal* **11**(13): 4767-4778.
386. Gao Y, Thomas JO, Chow RL, Lee GH, Cowan NJ (1992) A cytoplasmic chaperonin that catalyzes beta-actin folding. *Cell* **69**(6): 1043-1050.

387. Sternlicht H, Farr GW, Sternlicht ML, Driscoll JK, Willison K, et al. (1993) The t-complex polypeptide 1 complex is a chaperonin for tubulin and actin in vivo. *Proceedings of the National Academy of Sciences of the United States of America* **90**(20): 9422-9426.
388. Vinh DB, Drubin DG (1994) A yeast TCP-1-like protein is required for actin function in vivo. *Proceedings of the National Academy of Sciences of the United States of America* **91**(19): 9116-9120.
389. Kubota H, Hynes G, Carne A, Ashworth A, Willison K (1994) Identification of six Tcp-1-related genes encoding divergent subunits of the TCP-1-containing chaperonin. *Curr Biol* **4**(2): 89-99.
390. Lee MJ, Stephenson DA, Groves MJ, Sweeney MG, Davis MB, et al. (2003) Hereditary sensory neuropathy is caused by a mutation in the delta subunit of the cytosolic chaperonin-containing t-complex peptide-1 (Cct4) gene. *Human molecular genetics* **12**(15): 1917-1925.
391. Kubota H, Hynes G, Willison K (1995) The chaperonin containing t-complex polypeptide 1 (TCP-1). Multisubunit machinery assisting in protein folding and assembly in the eukaryotic cytosol. *European journal of biochemistry / FEBS* **230**(1): 3-16.
392. Delarue M (1995) Aminoacyl-tRNA synthetases. *Current opinion in structural biology* **5**(1): 48-55.
393. Fourmy D, Meinnel T, Mechulam Y, Blanquet S (1993) Mapping of the zinc binding domain of Escherichia coli methionyl-tRNA synthetase. *Journal of molecular biology* **231**(4): 1068-1077.
394. Xaplanteri MA, Papadopoulos G, Leontiadou F, Choli-Papadopoulou T, Kalpaxis DL (2007) The contribution of the zinc-finger motif to the function of Thermus thermophilus ribosomal protein S14. *Journal of molecular biology* **369**(2): 489-497.
395. Chiocchetti A, Zhou J, Zhu H, Karl T, Haubenreisser O, et al. (2007) Ribosomal proteins Rpl10 and Rps6 are potent regulators of yeast replicative life span. *Experimental gerontology* **42**(4): 275-286.
396. Hansen M, Taubert S, Crawford D, Libina N, Lee SJ, et al. (2007) Lifespan extension by conditions that inhibit translation in Caenorhabditis elegans. *Aging cell* **6**(1): 95-110.
397. Kaeberlein M, Kennedy BK (2007) Protein translation, 2007. *Aging cell* **6**(6): 731-734.
398. Cohen E, Bieschke J, Percivalle RM, Kelly JW, Dillin A (2006) Opposing activities protect against age-onset proteotoxicity. *Science (New York, NY* **313**(5793): 1604-1610.
399. Reinders J, Wagner K, Zahedi RP, Stojanovski D, Eyrich B, et al. (2007) Profiling phosphoproteins of yeast mitochondria reveals a role of phosphorylation in assembly of the ATP synthase. *Mol Cell Proteomics* **6**(11): 1896-1906.
400. Paul MF, Ackerman S, Yue J, Arselin G, Velours J, et al. (1994) Cloning of the yeast ATP3 gene coding for the gamma-subunit of F1 and characterization of atp3 mutants. *The Journal of biological chemistry* **269**(42): 26158-26164.
401. Hunte C, Palsdottir H, Trumpower BL (2003) Protonmotive pathways and mechanisms in the cytochrome bc1 complex. *FEBS letters* **545**(1): 39-46.
402. Martensson J, Meister A (1989) Mitochondrial damage in muscle occurs after marked depletion of glutathione and is prevented by giving glutathione monoester. *Proceedings of the National Academy of Sciences of the United States of America* **86**(2): 471-475.
403. Merad-Boudia M, Nicole A, Santiard-Baron D, Saille C, Ceballos-Picot I (1998) Mitochondrial impairment as an early event in the process of apoptosis induced by glutathione depletion in neuronal cells: relevance to Parkinson's disease. *Biochemical pharmacology* **56**(5): 645-655.
404. Amari F, Fettouche A, Samra MA, Kefalas P, Kampranis SC, et al. (2008) Antioxidant small molecules confer variable protection against oxidative damage in yeast mutants. *Journal of agricultural and food chemistry* **56**(24): 11740-11751.

405. Drakulic T, Temple MD, Guido R, Jarolim S, Breitenbach M, et al. (2005) Involvement of oxidative stress response genes in redox homeostasis, the level of reactive oxygen species, and ageing in Saccharomyces cerevisiae. *FEMS yeast research* **5**(12): 1215-1228.
406. Vitvitsky V, Mosharov E, Tritt M, Ataullakhanov F, Banerjee R (2003) Redox regulation of homocysteine-dependent glutathione synthesis. *Redox Rep* **8**(1): 57-63.
407. Cherest H, Thomas D, Surdin-Kerjan Y (1993) Cysteine biosynthesis in Saccharomyces cerevisiae occurs through the transsulfuration pathway which has been built up by enzyme recruitment. *Journal of bacteriology* **175**(17): 5366-5374.
408. During-Olsen L, Regenberg B, Gjermansen C, Kielland-Brandt MC, Hansen J (1999) Cysteine uptake by Saccharomyces cerevisiae is accomplished by multiple permeases. *Current genetics* **35**(6): 609-617.
409. Goldberg AA, Bourque SD, Kyryakov P, Gregg C, Boukh-Viner T, et al. (2009) Effect of calorie restriction on the metabolic history of chronologically aging yeast. *Experimental gerontology* **44**(9): 555-571.
410. Hochgrafe F, Mostertz J, Pother DC, Becher D, Helmann JD, et al. (2007) S-cysteinylation is a general mechanism for thiol protection of Bacillus subtilis proteins after oxidative stress. *The Journal of biological chemistry* **282**(36): 25981-25985.

I want morebooks!

Buy your books fast and straightforward online - at one of the world's fastest growing online book stores! Environmentally sound due to Print-on-Demand technologies.

Buy your books online at
www.get-morebooks.com

Kaufen Sie Ihre Bücher schnell und unkompliziert online – auf einer der am schnellsten wachsenden Buchhandelsplattformen weltweit!
Dank Print-On-Demand umwelt- und ressourcenschonend produziert.

Bücher schneller online kaufen
www.morebooks.de

OmniScriptum Marketing DEU GmbH
Heinrich-Böcking-Str. 6-8
D - 66121 Saarbrücken
Telefax: +49 681 93 81 567-9

info@omniscriptum.com
www.omniscriptum.com

Printed by Books on Demand GmbH, Norderstedt / Germany